Nothing But Blue

Nothing But Blue

A SUMMER AT SEA

Diane Meyer Lowman

SHE WRITES PRESS

Published November 13, 2018
Printed in the United States of America
Print ISBN: 978-1-63152-402-8
E-ISBN: 978-1-63152-417-2
Library of Congress Control Number: 2018942327

For information, address:
She Writes Press
1563 Solano Ave #546
Berkeley, CA 94707

Interior design by Tabitha Lahr

She Writes Press is a division of SparkPoint Studio, LLC.

Names and identifying characteristics have been changed to protect the privacy of certain individuals.

For Devon and Dustin, my best work ever. For my mom and dad, who I wish could have seen it.

"There comes a time in a man's life when he hears the call of the sea. If the man has a brain in his head, he will hang up the phone immediately."

—Dave Barry

Embarkation

June 1, 1979

40.685649N, 74.07154W

I walked up what I could only describe as a gangplank. A group of men in grimy coveralls hung over the ship's railing watching me ascend as they waited for the ashes from their cigarettes to fall. *What*, they must have been thinking, the *hell*? I could not even look back at my parents, as I'd had no idea what I was getting into until we arrived. By then it was too late to turn and run back to the safety of suburbia.

This German container ship would be my home for the next ten weeks. My cabin sat just below the bridge, right next to the captain's quarters. I later learned that I had unintentionally usurped it from one of the officers, probably the very one who had ushered us into it and then disappeared. The captain wanted me close to him. For my safety. It had more floor space than my dorm room. My parents flanked me as we tried to take it in, along with its implications. Small talk ensued as we walked around each other and tried out all the seating options like a triad of Goldilocks in an awkward game of musical chairs.

"The cabin is so big!" said my mom.

"I wonder what they'll have you do," said my dad. For as long as I could remember, he had me do "boy" stuff. The elder of two daughters, I was his "son." I mowed the lawn and raked leaves, embarrassed when my friends would ride by on their bikes on their way to town for Coke and fries at the Woolworth's counter, and see me, sweaty, hair back in a red bandana. He bought me Hot Wheels cars and took me to baseball games and taught me to keep score. Maybe that training would stand me in good stead in this sea of men, but I doubted that he saw the upside now as he considered the tasks this crew might assign me.

After some more strained conversation, there was not much to say to each other except goodbye, and we promised to write. We would not see each other for more than two months. I hugged them both very tightly, squeezing back tears. I did not want them to worry about me, but neither did I want to let go of my parents, my anchors. I'd only be able to receive mail periodically from the shipping agent in each port, and would be lucky if an occasional ship-to-shore call went through.

There was none of the fanfare associated with the good old days of luxury-liner sendoffs. No streamers or champagne. I did not wave from the deck, wrapped in a fox-trimmed coat, at relatives on the shore, excited to make friends with my fellow travelers over caviar in the dining room. In fact, it had begun to dawn on me that I had no idea what parallel universe I had chosen to enter. I walked my parents back to the gangplank and watched them descend, with the ghosts of the crewmembers hovering above me and murky, bottomless, black water below.

We smiled, and I tried not to cry. I tried even harder not to acknowledge the message in my father's eyes. I did not think this was what he had envisioned, or that he could conceive of leaving his nineteen-year-old daughter with these thirty-two German men in, what to him, must have seemed like a black hole, for ten weeks. But there wasn't a thing he could do about it. I could hardly bear to look at him. I read, *I am scared, I am sorry, I am worried* in his eyes. Like Peter Pan leaving Wendy with Captain Hook and the pirates, I watched them until I could no longer see them and then turned slowly to go back inside, hoping I could find my way back to my cabin. Eerie quiet followed me as I retraced my steps through the empty hallways. I passed only one young but haggard-looking woman heading out—I'd later learn that prostitutes came on board at every port—and locked the cabin door. Without a clue what I was supposed to be doing, I felt neither brave nor adventurous. I had just wanted to get the hell out of suburban Dodge for the summer, and hadn't given a great deal of thought to where I was headed instead. Testing out the blue coverlet on my new twin-sized bed, I stared out the large rectangular porthole, and wondered, *What's next?*

This had all started about a month before, when my dad had asked, "Do you want to work on a ship this summer?"

I stood at the end of the hallway at the pay phone, at the end of my sophomore year at Middlebury College, in the pastoral Green Mountains of Vermont. I called my parents collect every Sunday at the appointed time, just after I'd woken up. They declined the charges—our clever signal for *I'm awake and standing near the public phone blocking it so no one else can use it.* I'd hang up, feigning disappointment and indignation for the poor operator who had to go through this charade every Sunday around this time. My parents, on separate phone extensions, then called back.

"Do you want to work on a ship this summer?" my dad asked, as if he were inquiring about the weather.

"Excuse me?" I'd barely thought about summer. I supposed I would most likely work at Sight 'n Style, my uncle's eyeglass store in Brooklyn, where I'd worked prior to starting at Middlebury. Or maybe at Roy Rogers in my hometown of Westfield, New Jersey, where I'd donned polyester western garb to meet and greet fast-food aficionados during high school.

I immediately imagined myself as the cruise's social director in nautical short-shorts, limbs slathered with baby oil. I'd roam sun-filled decks encouraging happy passengers to play shuffleboard and consume large tropical drinks, and I'd emcee the belly flop contest. That would beat either the commute to Brooklyn or the disgusting smell of rancid french fry oil that I could never get out of my uniform.

"Sign me up! I'm on board!" Awful pun most sincerely intended. "When do I sail? Where does the ship go?"

Dad interrupted my reverie, telling me that Ted Williams, our neighbor, was the New York agent for a large German shipping company. He'd recently asked my dad if I might be interested in spending the summer on a ship. He had some pull with the company. They would let me work on board.

Okay! A German cruise ship. So, Europe?

"Are you kidding me? That's amazing! Where does it go?"

"Well," said Dad, sounding as if he has a special secret to share, "That's the interesting part!"

Interesting? Coming from a parent that usually means bad.

"Hamburg-Süd runs container ships."

Container? Like tropical drink containers?

"They ship cargo worldwide," he said.

To the Caribbean?

"You'd sail on one of their ships that go from New York to Australia and New Zealand and back, through the Panama Canal. It's a long trip; you'd leave right after school and probably get back just before you start again in the fall."

The cruise bubble burst. White sparkling beaches drifted away on puffy clouds along with the *biergartens* on the Rhine. But wait . . . Australia? New Zealand? G'day mate!

"That's awesome! But what is a container ship? What would I do?"

"Well, they ship freight. They dock at different ports to drop off and pick up cargo packed in large metal containers. As far as what you'd do. . . ." he trailed off.

His pause sounded like the non-verbal equivalent of *that's interesting*.

My mom hadn't said a word. That was hardly unusual. My dad's strong and controlling personality dominated. I was curious to know what she thought, but he left no room to ask.

"You'd be a workaway," he said.

"A what?"

"A workaway. It's a play on the term stowaway. Honestly, I have no idea exactly what you'd do, but I'm sure Mr. Williams can give you more detail." But as it turned out, he never did. I'm not sure he even knew what a workaway did. He just, very thoughtfully, believed it would beat a suburban New Jersey summer.

"It sounds like a real adventure," added my mom finally. She was never a worrier. She had an atypical philosophy for a Jewish Mother. She was not one to say, "Call me when you get there. . . ." She expected the best, and knew if the worst happened, she'd get a call. I admired and appreciated this. My dad had always wanted us to have experiences that growing up a first-generation son of Greek immigrants in the Bronx hadn't afforded him. He wanted me to go, and would make sure I did. He wanted to have this experience himself, but he couldn't. Consciously or not, he manipulated me with the slow revelation of the facts of the assignment; his unbridled enthusiasm leapt through the phone. I had no idea what a container ship was. I had no idea what kind of labor was involved. I had no better plans for the summer. So of course, I said, "Yes!"

I stayed there motionless for some time, looking around at my cabin and trying to digest it. I did not have a sense yet for the scale of the ship. When I boarded, I could see only a small portion of the bright, tomato-red side, the first-deck railing, and the entrance to the white, eight-story superstructure where the crew lived. I would only later come to appreciate the enormity of this big red monster—this sailing city. I had a cold and my period, and I felt crappy. I thought about curling up in a fetal position on the couch, but activity seemed a better way to dispel my mounting anxiety. So with nothing to do, and no idea when someone might come to give me an orientation—or even just a greeting—I tentatively explored my new environment and began to unpack my shorts and T-shirts, having no idea whether what I had brought was appropriate. Mr. Williams had given me no wardrobe advice, so I'd packed what I'd wear at home during the summer. I hoped the attempt to settle in would take my mind off my simmering panic. I felt like an interloper already. I did not want to venture out and bother anyone with questions, and besides, I knew neither whom nor what to ask.

My large cabin was impressive compared with the sardine cans that some of the sailors shared. I'd passed a few open doors on the lower decks, and what I glimpsed looked nothing like this. I took my shoes off and padded around the flat, marbled, golden carpet. Warmth radiated up from the floor, and I was thankful for the padding, knowing it would prevent sound from traveling to the cabin below. I wanted not to bother anyone.

Most of the furniture was built in so nothing would move when the ship did. I pulled a wooden armchair, its seat and back upholstered in the same blue as the bedspread and couch, away from the small table attached to the wall. Sitting down with my bag next to me, I took out a few family photos and my journal, and tried to imagine myself writing or sipping tea here while at sea. I missed my parents already, as if they were far away, even though we were still tethered to the dock.

I moved over to the couch and sat just where they had sat

as we'd chatted before they left, feeling their warmth, their imprints in the cushions. Light poured in from the porthole (which I later learned never, ever, to call a window) framed by golden drapes held open by a matching tieback. I took in the cabin from the couch for a moment, and then knelt on it with my elbows on the wide shelf of white space under the porthole. I could see nothing but imposing, multicolored stacks of containers, and a bit of the port beyond. A loud silence surrounded me. The stillness only heightened my anxiety. I had no idea what to expect, and only scary thoughts filled the silence. I got up and moved around to keep them at bay.

A tall, wide, built-in closet nestled in the wall space next to the couch, catty-corner from the door and opposite the bed. It had a rod for hanging clothes, as well as several drawers and a mirror inside the door. It reminded me of my dorm room closet, but bigger. I chuckled to think that a ship's officer, who presumably wore the same thing every day, needed more closet space than a college girl. I was uneasy about whether I'd be dressed appropriately for whatever I'd be doing on board. It didn't matter now, I realized, as I unpacked. There would be no way to buy anything else for weeks. For miles.

In the opposite corner, the twin mattress sat on a platform with more drawers underneath. Later I learned that the beds are always positioned perpendicular rather than parallel to the sides of the ship, so that when it rocked, I'd slide head to foot rather than rolling off. The nightstand had a small drawer and an open shelf on the bottom; a lamp attached to the wall just above it, and a beige rotary phone was affixed to the wall next to that. Who would I need to call? I was afraid I would have no one to talk to for months. I seriously doubted that anyone on the ship would be calling me, and I knew that no one from home could. It was only an illusionary lifeline.

I took out the books I'd brought to fill my time, stacked them carefully on the bottom shelf toward the back, and put the portable eight-track tape player in front to prevent them from sliding out. While arranging the tapes alphabetically—Genesis, Jethro Tull, Pink Floyd—I realized the plug was wrong. It hadn't occurred to us that a German ship would have European outlets, and I had no converter. This technological setback crushed me. My music would have been a thread of connection to home. I was a DJ at school, with a regular show on WRMC-FM. I'd picked those eight-tracks carefully; they linked me to people back home, and would free me, however temporarily, from the confinement of this cabin, this ship. I felt defeated before we'd even left the dock, and wondered where I might find a converter. In the Atlantic.

I walked into the "head," or bathroom, to the left of the cabin door, to unpack toiletries. I learned very quickly, after a few near falls, to step up over the three inches or so of wall at the bottom of every doorway. It would be the first of many things I had to avoid, sidestep, or maneuver around on board. The barrier prevented water from spreading throughout the ship, should the corridors flood. I tried not to think about that eventuality, and focused more on not tripping every time I took a step. Like looking left for traffic in the UK, this took a few near misses to master. The loo was as tight as the cabin was roomy, not much bigger than an airplane bathroom. I was thankful it was private and not down the hallway. The lower levels featured shared crew hall bathrooms. I was frightened enough of running into my shipmates fully clothed in broad daylight. I couldn't imagine bumping into anyone towel-wrapped after a shower. My own shower was snug, with a handheld showerhead that hung from a hook that I had to stretch to reach. Clearly most German officers were taller than my five foot two inches.

I made quick work of stowing what little I'd brought. Still, no one had contacted me, and I had no idea when someone might. I dared not venture out beyond my door, so I sat back on the bed where I'd started and hoped both that someone would come find me . . . and that no one would. Ever. I read *Time* magazine slowly, cover to cover. I could hardly focus; I read every sentence over and over, and still they didn't sink in.

I stared at that beige phone, as if waiting for a boyfriend to call, but with more dread than excitement. I unfolded, wound, and stared at the travel alarm my dad had given me for the trip, wrapped in red leather that matched the ship. Three hours passed before someone knocked on my door. I startled at the sound, shocked, even though I'd been awaiting it. I was afraid to open the door, but more afraid to sit there alone much longer.

"Fraulein Meyer? Alois here. Third officer," he said, as he offered his hand. It was his cabin I'd co-opted, but I was just as glad not to know that when he came to get me. I'd have wanted even more to disappear into the deep, blue-speckled linoleum floor in the hallway I stepped into.

"It is time for dinner," he said in heavily accented English. It became obvious that very few of those on board except for the officers—who had very little interest in communicating with me—spoke any English. Those who did spoke a very broken, very thickly tinged version that I would constantly have to strain to understand. I lamented that I had not learned German, and suspected that the Spanish and French I did know would be of no help. My mother had studied German, and I wished she were here with me to help me translate. I wished she were here with me.

"Yes, I'm Diane," I answered. Who else would I be? Was there another nineteen-year-old American girl who had stolen his spacious cabin on board?

"Good. Follow me please." I did. Like a geisha, five paces behind him, with my head down, studiously watching my own hands clasped together in front of me. We descended to the level I'd entered on. The crew's quarters and mess hall were down way below the officers' aerie near the bridge. The thick air down there smelled of heat, humidity, and cabbage.

The crew's dining area—the "mess"—was utilitarian. Beige linoleum floors held ten matching Formica-topped tables bolted to the floor. The chairs reminded me of the ones I'd grown up with in 1960s Howard Beach, Queens. Stiff vinyl seats and shoulder-blade-high backs, the color of butter left out on the counter. Chrome legs. Clearly not made for lingering over extended meals. The portholes were the only things that punctuated the beige walls. Cooking odors as thickly accented as Alois's English enveloped us.

When we stepped in, the chatter and clang of silverware went silent. All heads turned toward me. I already felt lightheaded from my stuffy head and cramped abdomen. I watched them watch me. *Please let me not pass out right here. Please let me not pass out.* Alois pointed me to the seat closest to the door, facing the others, and disappeared. He had dropped me in place like a discarded rag doll. And like a rag doll, I sunk limply into the chair. I had no context to make sense of any of this, so I just followed were I was led and took cues from others about what to do.

The show over for the moment, the men resumed eating. Soup steam soothed my sinuses, and I could just taste the warm bread. It was a harbinger of the freshly baked goods on board that would be my delicious downfall. And tea. Very distinctively flavored and ubiquitous green tea, always on the table in bronze plastic carafes with black flip-up lids. It eased the squeeze in my head a bit. The tea I sip now when I feel a cold coming on takes me right back to that table and the first moment I tasted it. I finished quickly and slunk back to my cabin, guideless, with eyes still averted lest I meet anyone along the way.

We were set to push back from the dock at 11:00 p.m. or 2300 hours; container ships use military time, and that was just one of the things I had to get used to quickly. But the seafaring life held many surprises and lessons for me. Like learning to step over those doorframes. Like the fact that arrival and departure times for container ships, unlike cruise ships, were merely broad, unreliable estimates. This big red vessel would challenge my control-freak, early-is-on-time tendencies for the duration, and stretch what was to have been a ten-week round trip to nearly twelve. I hoped we would depart while I was still awake so I could watch, but I gave in to exhaustion and crawled under the crisp white sheets to sleep. I checked the lock on the cabin door for the third time, and put one of the table chairs in front of it. To protect me from what, I was not exactly sure. At that moment, I feared everything.

The beige phone startled me just as I began to drift off a few moments later. In that hazy half sleep, I hoped against hope that my parents were calling on separate extensions like they did every Sunday at Middlebury. But sleep-induced delusion prevented me from recognizing that the phone was only a glorified intercom, a way for the crew to communicate on board. I groggily wondered if I could order room service on it. And laughed with myself at my little joke. Who in the world, or on the ship, could be calling me? When I'd stepped on board, I had become anonymous. I picked up the receiver with considerable trepidation.

"Herr Most here." It sounded more like "Must" in his mouth. "Chief steward. You will report to me while you are on board." I heard my father in the moment before I fully woke up.

"Yes, it's Diane." I had no idea what else to say.

"You come to the officers' mess. Meet me in the galley there. Oh eight hundred tomorrow." *Oh eight hundred?* "Breakfast first. Seven thirty. *Ya?*"

"Yes, I'll see you then."

"Good." *Goot,* I heard. He hung up.

I curled back up in a fetal position and waited to sleep. The ship, even when still, hummed with the generators' steady

Zen vibration. It seeped into my muscles and lulled me, but not for long. My stuffy head made breathing difficult, especially lying down, especially on pillows that were not mine. Cramps gripped my innards. Every time I heard a sound, I threw off the covers, leapt over to kneel on the sofa, and looked outside to see if we were pushing back, but there we stood, still.

Had I known what it would feel like when the engines started up and we began to move, the false alarms would not have fooled me. The ship, so wide that she just fit through the Panama Canal, and several football fields long, literally roared to life. I felt the unmistakable and unforgettable sensation in every cell. The quiet vibration of the ship at rest supercharged to a rattle evident in everything, including my bones, on board. Constantly. 24/7. Anything with any mass that came in contact with anything else clanged.

The ignition jolted me out of the tenuous deal I'd made with sleep at 0600 hours. Light poured through the portholes I'd failed to cover with the curtains. We peeled away from the dock with a motion I felt before I could even get to the porthole to see it. My stomach sunk and tightened. Although I'd driven away from home the day before, now I knew the leaving was for real.

I folded myself up into the wide windowsill, really more like a shelf that just fit me, to watch. It reminded me of the space behind the back seat of the VW Bug we had as kids. My sister Suzanne used to ride, all tucked up, in that small space between the upholstery and the engine during family drives. I thought of her as I hugged my knees to my chest in the smooth plastic space below the porthole, my back against the wall. She was three years younger than I, and off on her own adventure to live with a Navajo family on a reservation in Arizona. I would be at sea on her birthday. This was the first time in her sixteen years that we would not celebrate together.

The motion riveted me to the Plexiglas pane. The dock sat on my side of the ship, so I saw it recede, the lifted crane arms waving goodbye. After a few minutes, the open harbor approached. *Eyes forward. Don't look back.* The tugboats were invisible to me from this vantage point, high up in the superstructure. So tiny compared to the ship, they adeptly finessed this leviathan through the densely populated New York port and safely out to sea, gently guiding us with their thick rubber bumpers.

What I did see was the tip of Manhattan, the Twin Towers presiding proudly, looming large, but then diminishing. The Statue of Liberty waved goodbye, but then she began to disappear as well. Slowly at first, while we were still on the local streets of the inner harbor, and then more rapidly as we passed under the Verrazano-Narrows Bridge and out to the superhighway of the Atlantic. I had never seen a bridge from that vantage point before. It was fascinating, but imposing and intimidating. I felt relieved when we had cleared it. The protected harbor kept the surface calm and the waves at bay, but away from it, the water seemed more ominous. Slowly, the coastline disappeared, and there was nothing but blue.

At Sea

June 2, 1979

40.6066 N, 74.0447 W

Winged marionettes, the gulls followed the ship for quite a distance, scavenging for food scraps. Once this feathery halo vanished, land was a distant memory. We were at sea.

I could have stayed and gazed out for a long time. I would never tire of the intricacies of maneuvering my massive new home in and out of port, nor of the magnificence and ever-changing moods of the ocean and its *doppelgänger*, the sky. But I had to get down from my perch and break the reverie. It was time for breakfast, and I could not be late for my first day of work. Downstairs, I slid into the same seat I'd sat in at dinner. I had no idea if we had assigned places, but I didn't want to break a rule before I even knew if it existed.

"*Morgen*," said Claudia, the crew mess stewardess, one of only two other women on board. She nodded quietly as she placed a basket of warm breads, butter, and honey on the table.

"*Morgen, danke*," I said with a smile. I knew at least that much German.

"Tea for you, *ya?*"

"*Ya, bitte. Danke.*" I simply could not over-thank people on board. As on shore, I wanted to please everyone, starting with my father. But I felt it particularly acutely here.

She returned with a familiar plastic carafe of the vaguely eucalyptus-flavored green tea. As popular as green tea is now, it was virtually unknown in the States then. It was the only thing we got to drink at meals, unless we purchased soda or, like most of the crew, beer with our own money.

Claudia was tall, slim, and only a few years older than I. Her nose was slender and pointy, and she had straight teeth but a pronounced overbite. A crown of tight, dingy-blond curls piled high on top and shorter on the sides made her look a little like an exotic bird. She was permitted on board because she was married to the second cook, Bruno, who looked swarthy and threatening to me. He was as dark, compact, and tightly wound as she was light, lithe, and loose. In contrast, the first cook, Ingo, was as wide open as the sea we sailed. He was a big, hulking man with close-cropped blond hair, a little soul patch on his chin, and a laugh that filled the mess hall. If it weren't for his impish wit and a smile as big as his belly, he'd be intimidating too. His ubiquitous white apron tried, but could not hide his Santa Claus, beer-induced girth.

"*Morgen, Fraulein,*" he said. He smiled and winked, but not in a lewd way. He was the only one besides Claudia to greet me at all.

Again, I ate and left quickly, eager to report for duty. I felt like the train wreck everyone rubbernecks to gawk at with a mixture of curiosity and disgust. Once back in the hallway, I realized I had no idea where either the officers' mess or the galley were, so I slipped back in, and motioned meekly with my head to Claudia. I did not want her to think I was summoning her, nor did I want to draw anyone else's attention. But I needed her to point me in the right direction. Fortunately she saw me and came over quickly; I must have look distressed.

"*Ya?*"

"I'm sorry to bother you. Can you tell me, *bitte*, where is the officers' mess?"

"*Ya, ya.* Take the stairs to the top. Follow the hall to the wooden door."

"*Danke,*" I said, as if she had given me something much deeper than directions.

In the superstructure's internal stairway, the rarified air became lighter and sweeter as I neared the level of the bridge, the officers' area, and my cabin. Rank, I began to see, had its privileges. I pushed tentatively on the wooden, windowed door and stepped, like Dorothy, from drab into Technicolor. The internal hallway floors begrudged their muted blue, and the practical and practically colorless walls deliberately didn't shine. For a moment I straddled both worlds, one foot decidedly part of the lower caste while the other floated slightly above the deep pile of the officers' deeper-than-ocean-blue plush. The eddy of affluence pulled me in. The reverent and respectful quiet did not allow for the *clip-clop* of shoe soles up here. The long, polished-to-gleaming wood table and amply upholstered chairs whispered rather than rattled. The china and crystal clinked only politely, unlike the plebian clatter of the crew's plastic dinnerware. The room was soft-focus backlit, in stark contrast with the fluorescent glare downstairs. Carefully curated and appointed nautically themed art adorned the paneled walls.

My head revolved in slow motion to take it all in, including the few crisp, white-shirted, black-trousered officers lingering languidly over their morning meal. Unlike the crew downstairs, the officers barely looked up. They could not be bothered; I was not that important to them. Then my gaze met Herr Most's. Waiting sylph-like, blending into a corner of burled wall, he was still and silent except for his beady blue eyes. They now spit sparks in my direction, as he discreetly twisted his head, as I had moments ago with Claudia, magnetically drawing me toward the galley door and away from the idyllic scene. I tiptoed over the twisted-wool carpet fibers and followed him as he slipped in.

He turned to me abruptly, and by way of greeting and introduction whispered sharply, "*Bitte*, do not use that door again, Fraulein Meyer. You come in here always, *ya*?" And he guided me by the elbow toward the separate galley door. "*Ya, ya*," I agreed. I'd made a tremendous but unintentional faux pas by entering directly into the officers' mess, and probably embarrassed him. Not a particularly auspicious start to our working relationship.

I couldn't do anything but hang my head and nod. Herr Most was not much taller than I, with a slight, tight, wiry build. I never saw him wear anything but the uniform he had on then: a starched-stiff, white, short-sleeved shirt tucked into slightly bell-bottomed, black trousers, with a matching belt and utilitarian, thick-soled shoes.

His closely cropped, gray hair framed fine, chiseled features. There was not an ounce of fat on his frame, and I immediately sensed it was because he was in constant motion, even when he was standing still. I never saw him sit and could not imagine him prone, even to sleep. His hands were always busy; he looked like he was fingering a cigarette even when he wasn't "schmoking." As chief steward, he had many responsibilities that mostly included keeping the officers happy and the crew under control. And now, figuring out something for me to do.

After a quick tour of the long, narrow, mostly stainless-steel galley, he broadly outlined meal schedules, including "schmoke" time, at 1500 hours. Similar to teatime in England, on board it included cigarettes in addition to hot beverages and warm, freshly baked pastries. Almost everyone on board "schmoked." My non-partaking of the tobacco ritual set me yet further apart from those on board. Given the fact that fire was one of the most dangerous threats on a ship, the habit seemed ill-advised, but I dared not venture that opinion.

He led me through another door to a flaxen-carpeted and upholstered lounge, outfitted with a television and radio, which only functioned when we were within antenna range of a port. He pointed for me to sit—I was dropped like a ragdoll again—on a couch under one of the large portholes. Since we were so high up in the superstructure, the light that poured in gave everything a Golden Fleece hue.

On the cushion next to me sat an immense pile of fabric and a box of sewing supplies. I could not quite discern what the pile contained, but he pulled out a few items and said, "You fix." Some were fitted bed sacks that served as bottom sheets. They had tie closures, many loose, and some missing. I was meant to mend or replace those. Also, there were the ubiquitous coveralls that the crew wore while working: rough, heavy material that, despite its evident durability, had worn bare in spots. This testimony to work so rough that the fabric didn't stand a chance meant I would need to patch, darn, and mend where I could. If a pair was beyond repair, I would cut them up as fodder for future patches. Then there were more delicate officers' linens that I would need to finesse back to usability.

So. I would sew at sea. Betsy Ross of the *TS Columbus Australia*, representing our nation. I remembered the Home Economics course I'd taken at Westfield High School that had seemed so stupid back then. Clearly Herr Most assumed women knew how to sew from birth. I was now glad for that dumb class. It seemed a slow way to pass a long time, but the room was comfortable, the work easy, and I'd be left alone. I wondered if he had other things in store for me, but dared not ask.

As if he were reading my mind, he added, "And before meals, you will set up the officers' mess, *ya*?" He may have formulated it as a question, but I had no doubt that it was a directive. I would set their tables for lunch and dinner, but disappear before they appeared. He would clear the dishes because he did not want me in the dining room when the officers were there. He could not trust me with them yet.

The pile seemed endless, and surely Herr Most would replenish it regularly, but it was cool and bright in the lounge, and I could get up for a break whenever I liked. He showed me where I could help myself to coffee, tea, or water any time. My fingertips would bleed, heal, and callous over from the needlework, but over all, the work was hardly arduous. For now.

––––––––––––

While we still paralleled the Eastern Seaboard, the radio played familiar music, albeit on stations that changed as we sailed in and out of each city's reach. It was like constantly needing to tune and re-tune the car radio on a long road trip in our black VW Bug, except no pavement supported us. This fact uprooted my sense of balance and unsettled me, especially for the first few days. I felt the unmistakable forward motion of the ship through the water—which, out here in the Atlantic showed more muscle than it had in port. A churlish blue-green, it poked at the vessel like an annoying sibling, just to remind us it was there, but was no match for our bulk.

When the engines were on full power, as they were now, out in open water, the loud white noise formed the canvas on which everything else appeared. Its inescapable hum could be an *om* or a growl, depending on exactly where in the ship I was and what I was trying to do. And everything moved. Always. Even the furniture, even the heavy things that were bolted down vibrated. Smaller items—those merely bracketed or braced or cushioned—shook, slid, and clunked constantly, the cacophony changing with the ocean's mood. Smooth seas muted the inventory's symphony. When the waves acted up, it could sound like a bowling alley with everyone scoring simultaneous strikes.

As a result, we had to secure everything in some way: dishes, glasses, and cutlery sat, penned in by partitions, on spongy cushions. Books, as I learned quickly after I left a few on my small table, had to be corralled like milk bottles in wooden crates. Clothes in the closet swayed with the swell, and the hangers, imprisoned wind chimes, clinked together constantly. Shoes ended up in a pile in one corner of the closet, heedless of the nonslip surface they stood on. Toiletries tucked tightly into the medicine cabinet tried to tumble out whenever I cracked the door open to retrieve something.

Our bodies fell under the ocean's swell as well. I learned to tune in to my center of gravity, like the passing radio signals, at all times. Like parents of newborns who develop a rhythmic bounce with their babes in arms, I intuitively adjusted my body in space to accommodate the ever-shifting surfaces on which I walked, showered, and supped. I never felt steady. Of course, it was impossible to recalibrate while asleep, so I vibrated, rolled, and slipped all night in concert with the ship's speed and the ocean's sway.

Not only did I get used to this within the first few days, as the seasoned seadogs around me had, but I came to find stillness and silence more disturbing. On board, slowing signaled either a problem with the ship, or that we were nearing a port. The rattle and hum meant all was well. On land, it became hard for my body to readjust to solid, un-shifting surfaces. It still wanted to bob and duck like a prizefighter, and when my body moved to adjust to ground that wasn't in motion, it was very disconcerting to my equilibrium. I understood "sea legs" as a very real phenomenon; throughout the trip I felt more nauseous on dry land than on even the roughest day at sea.

I felt the downshift immediately, as I tried to corral the sewing supplies so needles and spools didn't escape into the cushions or onto the floor. Although the ship may have looked like an arrogant, inscrutable behemoth, it was actually very easy to read her moods and moves. I knelt on the tweedy mustard cushions and hoisted myself up so I could see out the porthole, although my body knew before my eyes did that we'd veered ever so slightly right. We'd been within eyeshot of the coast since leaving New York. I could see a thin crayon line of muddled brown just on the horizon, dividing the blue, dark for the sea and light for the sky. Now it was growing wider, and the V-shaped frothy wake the bow cut began to subside. We'd left New York less than a day before, and we were heading into a new port.

Herr Most came in to confirm what I already knew.

"We are near Philadelphia. You finish. You eat. You can watch, *ya*?"

I popped up, perhaps a tad too quickly, and composed myself to show a little less enthusiasm and a little more care as I stowed the supplies. The pile I'd been working on looked untouched and dwarfed the finished pile, but it wasn't a bad morning's work in my mind. Herr Most stood by, stoic and inscrutable.

I motioned to the completed work and started to ask what he wanted me to do with it, but he cut me off before I started. "Leave it. You come back the same time tomorrow morning." He used, as always, the imperative. I was never quite sure if he meant it to sound that way or if it was the limit of his facility with the language, but he always spoke to me in commands. "Through the galley door, *ya*?" he said. Like Cinderella. Like the help. Which is exactly what I was.

That was it? Lunchtime, and I was done? I made my way back down to the crew mess for the midday meal, with no clue what to do for the rest of the day.

I left the ivory tower of the ship's aristocracy and headed back down to dine with the common folk. But the problem was that I belonged to neither group. I was at once completely contained and entirely adrift. The lightness I felt as I finished work for the morning disappeared as I wound my way down the stairs, the air and my mood more sluggish with each step. I reluctantly slunk back into my seat.

Why, I imagined them whispering, *is she on board?* Every time I walked into a room or out of a room or by a room, I'd sense a slight pause in whatever everyone was doing, and hear the same refrain in my mind. *Why are you here? Do you think this is some kind of amusement park? You may be on vacation, but we are working. Hard. We are not amused, you entitled little bitch.*

And the whispers would grow louder, and the crew would grow in stature. In some kind of Alice in Wonderland, topsy-turvy, twisted-proportion distortion. I could barely communicate with the only two other women on board: Claudia and Ana, the diminutive Brazilian wife of one of the greasers, who seemed to be more like a souvenir from a port than a real person.

I'm sorry, I wanted to reply to their unspoken words. *I'm sorry that I am a young woman straight out of the American upper-middle-class family sitcom you are imagining. I'm sorry I'm here. Invading your space. Your lives. Upsetting the delicate balance on board.*

Eventually the crew and I would have the opportunity to ask, and answer, these and many other questions in person, trying hard to understand each other in the absence of a common language. Their English was broken, at best, except for the officers and a few of the crew. My German was tourist rudimentary, making it hard to clear up misunderstandings and misconceptions without a shared lexicon. It's one thing to try to make yourself understood when asking for the nearest focaccia bakery in Italy on vacation, but the struggle takes on a very different hue in a dark, below-the-sea-line, shared cabin of three of the lowest caste members in the evening after they've had several Holsten beers. But no matter what the time of day, or the rank of the sailor, or the point in our journey, the question was always the same: *What are you doing here?* And no matter what my answer, they never seemed quite satisfied.

So I ate quickly. My discomfort propelled me from the room. I could feel the ship slow and veer and tug me to see what would happen in port. I went out onto one of the lower decks to stay out of the way and out of sight, especially of the officers. Although we were still in the United States and only a short way from home, I really had no concept of things that had become so customary to the sailors. We slowed to what felt like a doggie-paddle as we got closer and closer in, and the waters around us calmed and grew more congested with other vessels, large and small. Ships our size didn't turn quickly, especially at slow speeds, so getting in and out of port safely took great skill.

Buildings came into sharper focus, and when we got very close, tactful tugs pulled up alongside us with their huge rubber bumpers to gently coerce the ship into its berth. I would develop a real affection for these mighty midgets, usually painted cartoon-brightly so as to be readily visible. Their crews knew every nook and cranny of their own port as intimately as melted butter knows its English muffin. In some larger or busier ports, pilots from the tugs actually boarded the ship like friendly pirates completing a bloodless, albeit temporary coup. The captain, who was solely responsible for the crew and all the cargo on board, maintained an iron fist of control over his ship. It was only at these times that he would voluntarily let an outsider wrest control of his charge so willingly.

I waved at the rough-looking, overall-clad, burly, five-o'clock-shadowed men on the tugboat decks, and they always smiled and returned the greeting. They treated our big red beast of a ship with the tenderness a mother reserves for her newborn.

Back then, we were one of the first generation of ships to be outfitted with thrusters that allowed some lateral movement, but even these were not sufficient to guide us into a slip safely. So those aquatic assistants nudged us until we were lined up parallel to and within feet of the concrete dock. I watched the crew unfurl heavy, twisted, metal cables from the decks and hook them over what looked like huge steel chess pawns to secure us in place. This was not a procedure for the impatient. Not like my father who had bequeathed to me his ability to one-handedly whip his car backward into a parking spot. The whole operation, from entering the mouth of the port to tethering, could take hours, and varied with the size and traffic of both the port and the specific dock. It was a thing of beauty—a ballet, really—danced silently except for the tender toot of the tugs and the bass-y bellow of our ship's unique, UFO-shaped funnel.

Nothing happened quickly on the ship, and that's something to which I had to inure myself. I'd inherited both impatience and efficiency from my father, who would do whatever he could to control circumstances and circumnavigate rules to get what he wanted when he wanted it. It has taken me a lifetime to recognize the nuclear fallout of this type of behavior, and I work hard to modify it. The seeds of change were planted on board that summer. Almost everything important to a ship's smooth operation is beyond its control: weather, tides, and dock operations. They could anticipate and plan as much as possible, but things changed, always.

The ship may have sidled up and been secured, but unless myriad workers were available and conditions were met, we would just sit and wait—and often did. But things went well in Philadelphia.

I don't know the exact capacity of the *TS Columbus Australia*, and she no longer sails for Hamburg Süd, but their *Spirit of Sydney* has a capacity of 3,630 "TEU" (twenty-foot equivalent unit). Empty, a twenty-foot container weighs around 5,000 pounds; full they may weigh 65,000. Math may not be my forte, but any way you look at it, that's a lot of weight to haul around.

Enormous cranes (called gantry cranes) stood sentry at attention waiting for us to dock, their arms (called spreaders) almost as long as their bases, pointed straight up initially, in the pose of a diver poised on the high-dive platform. The operator lowered them parallel to the ground, creating a ninety-degree angle.

I watched the cranes standing over two hundred feet high, with almost that wingspan, travel over the full width of the ship as they unloaded and reloaded the containers waiting their turn patiently. Their facility in sliding back and forth along the dock from fore to aft, making quick work of moving the weighty freight, awed me.

The crane operator slid the arm over each container like the claw in an arcade game that skims tantalizingly over the stuffed animals and treasures in slippery plastic capsules. He artfully centered the spreader over its target, and lowered the boom onto the top of the container. Unlike the arcade claws designed to miss, these cranes sunk their talon hooks into each of the four large holes in each corner, locked them in place, and lifted the enormous containers, making it look as easy as if they were lifting a stuffed panda and not a twenty-foot, fully loaded, metal box.

I stayed in place to watch them stack the multicolored containers into a Lego-like mosaic. The shipping line had its own signature red containers with Hamburg-Süd emblazoned on the side in white, but we carried all types, including the orange ones with blue lettering of our competitor, Hapag Lloyd. Some were refrigerated (for example, the lamb that we'd carry back from New Zealand), and others not. Yet when I asked one of the officers what was in each container, he shrugged. "Who knows? We just know which to load and unload at each port."

As the operator caught and carried off each container, he'd reel it back onto land like a sport fisherman who'd hooked a prize, and place it delicately on the dock to await its next destination. From there, some were placed onto waiting freight trains or tractor-trailers, or unloaded right at the port. The process, complex to me, went as smoothly as a well-rehearsed symphony. I imagined that crane lifting me off the ship and placing me carefully on the dock so I could call my parents to come get me in Philadelphia: "Hi, Dad, I changed my mind. . . ."

Charleston went as quickly as Philadelphia. We had little cargo for those ports. I did not venture out. There was really no time, I was afraid to ask Herr Most if I could leave the ship anyway, and I needed to get used to the routine on board. As nervous as I was, I wanted to get on with the adventure.

Container ships unload and load more quickly than other ships, so depending on how much tonnage needed to be moved around, we could be in and out in the nautical blink of an eye.

I kept mostly to myself those first few days on board. I did not want to come across as aloof or unsociable, but my cabin held me safely, and I had no idea what interaction would be proper and acceptable with the men. I couldn't communicate easily because of the language gap. At meals, certain groups always sat at the same tables, like in a middle-school cafeteria. I dared not break with this order, and preferred my seat right next to the door. It afforded easy egress. Karl, a quiet electrician, and apparently an unknown twin of Harpo Marx's, always sat across from me. He had loose blond curls and an easy manor that matched his smile. He would nod and share the time-appropriate, abbreviated greeting: "*Morgen*," or "*Abend*," leaving off the "*Guten*," and then sit down and eat quietly.

Alois, the third officer whose cabin I inhabited, and who had escorted me down to my first meal on board, made an effort to be friendly. With ulterior motives, I suspected. He brought to mind some Tolkien character with his solid build and wavy red hair. With a pointy hat and pointy ears, he'd have made an excellent garden gnome. His air was light and friendly, but I always sensed that he was trying just a little too hard. To fit in (as a low-level officer he shared my limbo between the grunts and the gods). To make the right joke. To add the right quip. His tone and timing were always just a little bit off. But his English was good, and so far he was one of the only people on board who had spoken to me.

"*Morgen*, Fraulein Meyer," he said.

I looked up from lunch, wondering why he was down here slumming.

"*So*. We have here on board every week on Wednesday movie night. We call it *Kino*. It means cinema. At 2000 hours. In the crew lounge. You come?"

I considered this only briefly. I usually finished work and dinner by 1900 hours, and the nights alone in my cabin, even this early on, stretched out like the ocean ahead of me. I longed to find a way to fill the hours and felt cautiously interested in interacting more with the crew.

"*Ya, danke.*" He explained how to find the lounge. I listened carefully so I could remember.

I sat in my room after dinner on Wednesday, on the couch, feet on the ground, palms facing down by my side, staring at the door. A strange hybrid of a deb anticipating a cotillion and a criminal awaiting trial. Six days into the trip, and I'd have both my coming out party and my first day in court. I didn't know if I should dress up and wished I could calm down. Too fidgety to read, I turned and kneeled on the couch cushion, arms resting on the space below the porthole, and watched the inkwell of endless ocean go by.

I decided to wear the exact same shorts and T-shirt I'd had on all day, with my hair back in a ponytail and no makeup. I did not want to look as if I'd primped in order to attract attention. I closed the door behind me at 1945 hours and left my solitary days on board behind, at least for the time being.

The lounge was dark, thank goodness. I slunk in and sank into one of the blue upholstered chairs, trying to make myself small. Alois came in and right to me.

"Ah, so you are here, *Fraulein*! Good. I will introduce you to some of the crew."

"This is the radio officer, Herr Stuhlemmer"

I smiled and shook hands with each like a Von Trapp child. I all but curtsied, and made only brief eye contact. I wanted to be polite, but I was intimidated and overwhelmed and felt as if I were on display for assessment. Glad when the lights began to dim, I curled up into a blue barrel chair near the door, as the attention thankfully shifted from me to the screen. And there, dubbed in German, was Bruce Lee. *Enter the Dragon? Fists of Fury?* I can't remember, because mostly I was dazed by the surreal combination. I wanted to call someone—anyone—at home. *You'll never believe what I just watched*, I'd say. But that couldn't happen. I was content to cocoon in the cool dark, lit only by the flickering screen, the men reacting to the film the soundtrack behind me.

When the credits rolled and the lights came on, I made to leave, afraid to loiter. I hugged the wall on the way out, as if I could fade into it and escape unnoticed, but someone grabbed my arm just before I reached the door. I started.

"You liked it?" asked Alois. "You come to have a drink with us now?" He motioned over his shoulder vaguely toward a gathering group.

"No, *danke*," I said. "I have to get up early to work, and I'm so tired and. . . ." I ran out of excuses and wanted to run out of the room.

"Another night, maybe?" He said.

"Yes. Thank you for inviting me," I said, and turned to seek sanctuary in my cabin, to let the tension that was twisting me unwind.

Movie night broke the ice and allowed for a slow thaw. Some of the men I'd met would come up to reintroduce themselves and ask a few questions or suggest an activity. I was still reluctant to dive into the deep end of this pool, but I appreciated the offers and assured each one that I'd join them as soon as I got my sea legs. But that joke was wearing thin, and I felt as if I were treading water on a fine line between reticence and arrogance.

We left the Atlantic Seaboard for the Caribbean en route to the Panama Canal. In the open waters, without the wall of land beside us before, the sky dawned brighter, the sea shone lighter, and the humidity hung heavier.

Radio Officer Herr Stuhlemmer summoned me to his office to discuss a few business items. Nerdy and a little pudgy, his white dress shirt strained to close over the belly that struggled to escape his khaki pants' waistband. His thin, mouse-brown hair matched his mousy voice. When he spoke, I could see little Chiclet teeth that didn't quite meet, and his cheeks puffed out just enough for me to suspect that he had some cheese saved up in them for later.

"You are enjoying your time on board?" he asked. It came out as a peremptory courtesy more than genuine interest.

"*Ya, ya,*" I said. What else could I say? In response to questions, *ya* always came in pairs.

"We have a safe in my office. If you want, you give your money to me and I will record the amount and keep it securely for you. It will be better here than in your cabin. I can give it to you when you need it. I can change money for you in Australia and New Zealand, but a bank will make a better exchange rate for you. You may buy items at *Kanteen* on Tuesday afternoon or Friday evening. These items will be charged to your account. The Chinese man can do your laundry. It will be charged to your account. You will leave your bed linens and towels outside your door with your laundry on Monday morning by 0800 hours; it will be returned to you in the afternoon. You can make ship-to-shore calls from here. The connection is not always so good. You will let me know in advance when you want to make these calls. They will be charged to your account. Do you have any questions?" He'd barely taken a breath. I hadn't.

My head was spinning. I wished I'd brought some paper to write all of this down. Who might go into my cabin and steal my money? Which morning was *Kanteen*?

"Um . . . no. Thank you. I will bring my money up, for sure. But wait, I do have one thing, if you wouldn't mind."

"*Ya?*"

"I brought an eight-track tape player with me. But I forgot a converter."

"Oh, *ya, ya,*" he said, more relaxed now that business was over. "I can find that for you, or one of the electricians can make it fix. Bring it up when you bring your money. There will be no problem."

"Thank you so much." Music would fill my cabin with so much more than melody.

He just straightened up and nodded, dismissing me. He had work to do.

On the way out of his office near the bridge, First Officer Rose stopped me and introduced himself.

"You are Fraulein Meyer, *ya?* Herr Rose here, if I can be of any assistance. Perhaps you will like to visit the bridge at night to see how we navigate by the stars."

"That would be so cool. I would enjoy that very much," I answered quickly because I so wanted to see those stars. If there was anything behind his invitation, I missed it. The officers were much more formal and disciplined than the crew. I looked at the wedding ring on his right hand, and while those didn't always assure purity of intentions, with him, I felt comfortable.

"Come up any evening that I have overnight duty." I wondered how I would find that out, but just thanked him and left.

But apparently Herr Rose and I had business to conduct before pleasure. The next morning, mid-stitch, Herr Most marched into the lounge.

"You go meet First Officer Rose on the Lifeboat Deck. *Schnell, ya?*"

"*Ya, ya.* Where is the lifeboat deck?" I asked.

He rolled his eyes, exasperated, as if he'd explained this to me a hundred times before. As if I'd been born on the ship. He explained quickly and I listened carefully. And again he said, "*Schnell!*"

Herr Rose awaited me in his dress uniform just past the heavy steel door that led from the superstructure to the deck. I felt like I should stand at attention and salute.

"*Morgen*," he said. "I must show you your lifeboat and explain the procedure for muster. We expect not to need this, but you must know."

I nodded. We stood on what was, to me, the right side of the ship.

"First," he began, "you must call the direction on board properly. The front. It is 'fore.' The back. It is 'aft.' The right, where we are now, *ya*? It is 'starboard.' And the other side, the left. It is 'port.'"

I nodded again. My mind worked furiously to remember these terms as it also wondered how I'd recognize them in German. I did not want to have to be told twice.

"Now, come." We walked a ways on the deck and stopped underneath a large rowboat suspended from pulley-operated cables above our heads. He looked around for a moment.

"The ship, she—is always female, the ship, *ya*?—she is very sturdy, but just in case, if you hear the alarm, you muster. You come together here, under this lifeboat. You take the life vest, here." He pointed to a white bench, whose seat lifted to reveal a bin full of orange puff. "Put it on and wait, *ya*? Quickly."

I nodded. "Good," he said. The word always sounded like "goot" in their mouths. "Fire is the worst risk for a ship. Nowhere to go. If you smoke (schmoke), you put them out only in these red buckets filled with sand. Do you smoke?"

I shook my head, no, wondering again why they let the sailors smoke on board.

He relaxed, having completed his duty. He was tall and trim, with dark hair and a thick, dark mustache. His deep, dark eyes made direct contact with mine, which was unusual. Most people so far looked down or up or anywhere but directly at me when they spoke. His direct gaze would have unsettled me if not for his bright, warm smile. His age and rank were far enough above mine that it said friend, not predator.

We leaned over the railing and watched the blue, as far as the eye could see, only changing texture when the sea became sky. The sea was friendly but playful, little white-caps cresting and disappearing like mischievous prairie dogs. Looking straight down, watching the red metal slice through, it seemed we were traveling very fast. The cut was precise and virtually silent. The watery siren sea hypnotized me.

"She has a top speed of twenty-two knots," he said. "Not so fast, but she carries a heavy weight. She works very hard every day."

I looked at her, this ruby giant, with new respect. Immense and heavily laden in an absolute sense, yet miniscule and seemingly weightless within the context of the even vaster field she plowed through. Regally, without complaint.

Herr Rose pointed his finger out toward the horizon. "You see the seagulls, *ya*? That is how you know we are near land. They follow her looking for food. If you look down, you will see sometimes dolphins or flying fish. They swim near the hull and leap out of the water with their wings spread. It is quite amusing."

"Wait, land?" I asked. "I thought we were well away from the East Coast and on our way to the Panama Canal."

"*Ya, ya*," he said. "But we pass islands: San Salvador, which your Christopher Columbus made famous." *My* Christopher Columbus? He had no edge to his voice, but others would. This would be the first time of many, on board and off, that I would assume full responsibility for and represent all things American—good, but more often bad. Historically, politically, economically, and culturally. It was quite a burden for a nineteen-year-old college junior.

"And Cuba. We pass Cuba, too. The gulls, they come quite far out. They are a harbinger of land."

I nodded, and thought, *harbinger*? So he was smart, too.

"We will make the canal in another day, although no one can say when we will go through. We wait just outside, anchored with all the other ships for our turn. You will see many different kinds of ships there: car carriers, they are immense; they look like big, floating tanks. Cargo ships with their own cranes that carry their payload on pallets in their hold. Long oil tankers that have yards and yards of pipe on deck. You can tell how full a ship is by looking where the hull meets the water. If full, you will see only a smooth hull line: one color. If emptier, the fore part of the hull will rise up and you will see the indent of the keel. Usually painted a different color."

I nodded. This was more than anyone else had taken the time to explain.

"Have you met Herr Kapitän Beucking yet?" he asked.

I shook my head.

"You will. He will relax more once we are through the canal. Several of us take responsibility for the bridge. We rotate eight-hour shifts." He looked around, his hands clasped behind his back, hair too short to blow in the strong breeze, as if trying to remember if he'd forgotten anything. Official again, he said, "Do you have any questions?"

"No. Thank you *so* much."

I wanted to hug him. He was the first person on board who had treated me like who I felt I was. But he had already turned crisply and disappeared back inside.

Point of No Return

June 11, 1979

9.3593 N, 79.8999 W

The ship slowed, and the pull on my body ebbed. I dropped the top sheet and the needle and thread and knelt on the couch to look out. A crescent of land—looming dark from this distance—was barely discernable where it met the now darker and more still water that held dozens of illuminated anchored ships. I left the linens where they lay and bounded into the galley like a puppy that wants to go out.

"Are we there?" I asked Herr Most, who stood with his back to me, repositioning some glasses on the nonslip cushion in the cabinet. He didn't turn around.

"*Ya, ya.* Soon. You make finish and you can go look on deck. But we won't go in today. We will wait."

I hated to wait. But I didn't let on. He looked disappointed that his comment about waiting hadn't riled me. He already knew I was always anxious to get going, to move forward, and had already begun to push back by needling me about delays. The schedule, as printed, did not leave much leeway between our projected return date and my first day of school. Herr Rose had told me that we'd likely anchor awaiting our turn, so I was armed with forewarning.

"It could be a long time," he added, looking down at his hands and not directly at me. "You never know. It depends on the queue of other ships." Now he looked up to see the impact of his words.

"Okay," I said. "Do you need me later?"

He frowned and looked down at his nails, stymied for a moment, either wondering why I'd taken the news so casually or trying to think up something for me to do later. I tried not to bounce and fidget as I waited.

"You clean up the mending before you go, *ya*? Then make finish for the day."

"If we go in tomorrow can I watch? Do I have to work?" I practically wagged my tail with excitement.

"You come here first, and we see what happens."

I darted into the lounge to tidy up, not wanting to miss a minute of our approach, feeling like he'd already kept me too long. I stowed the supplies and stumbled out on deck.

The land came toward us. We, the camera, shifted focus from a long shot to a close up. With each passing moment, palm fronds crisped, ship outlines solidified, and ours shrunk in comparison to the others. She was enormous and solitary from the on-board-alone-at-sea perspective, but she normalized relative to the other vessels waiting to "go in." She was just another ship.

We dropped anchor and settled into our spot in the canal's holding basin. I popped back and forth like a ping-pong ball from port to starboard through the superstructure. From one gulp of heavy, hot, red air on deck, through the cool, blue linoleum air inside. Back and forth, back and forth, eager to check out all the neighboring ships, and loathe to miss anything. If any of the crew could see me, I'd confirm what they already thought they knew: I was crazy.

An overwhelming number and variety of fellow vessels

dotted the water as far as I could see; too many to take in and process at once. This commercial navy laid out like a big metal blanket over the still cove filled me with a deep sense of connection to these anonymous seafaring comrades. No matter what their country of origin, what they carried, or where they were headed, we shared something very fundamental. We formed a community of itinerant transporters. I felt a deep pride for the *TS Columbus Australia*. She was the prettiest, most regal ship around. Her unique white, round funnel crowned her bright red body such that she stood out among the other, mostly black, many dingier boats with their mundane rectangular funnels. I could see signs of wear and rust on some of the nearby oil tankers, and the gargantuan car carriers had no personality. The cargo ships looked wimpy compared to our champion weight lifter; she was as large a ship as could safely pass through the canal. In this gathering of dinosaurs at the watering hole, we were the Tyrannosaurus rex.

It was a UN General Assembly of countries as well. Every ship's registry was writ large on the hull for all to see, and each proudly flew the flags of its mother country (often different from the ship's registry) and its parent company. In port, the ships also flew that country's flag. So the masts fluttered with a multicolored display of allegiances.

Sated with this scenario, I went back inside, hoping to find someone who knew when we'd go in. But no one on board ever felt the need to tell me anything, and I would not have dared approach the bridge to ask. I hoped I'd run into Alois, because he might know. It would give him great pleasure to have something to hang over me. I couldn't ask Herr Most because he'd just dig his thumb into the paper cut of my impatience. With no one in sight—everyone else was working or sleeping, in the case of overnight shift workers—I finally admitted defeat and went back to my cabin to collect most of my cash and the tape player to deliver them to Herr Stuhlemmer, with the very in-plain-sight ulterior motive of seeing if he might have any inside information for me.

"Welcome, Fraulein Meyer." I kept encouraging everyone to call me Diane, but the confluence of German and ship etiquette seemed to strictly prohibit it.

I handed him the money and the eight-track player. "I make you a receipt for the money. When you need it, you advise me and bring me this receipt and I will mark the withdrawal. You leave the player with me. We will find a way to make it work."

"*Danke*," I said, as he turned back to his desk, busy with paperwork for our impending passage through the canal.

"Herr Stuhlemmer," I began, and he turned back toward me.

"*Ya, ya?*"

"Do you have any idea—"

"When we will go in?" he interrupted. "*Nein*. It is hard to say. It depends on the other ships waiting, our place, our size—many factors—maybe tomorrow, but we have waited longer before." He looked at me for a moment and turned back to his desk. "*Guten Abend, Fraulein.*"

I headed down to dive into *Airport* to pass the time until we began to move. The irony of reading a book about air travel while making my way around the globe at a comparative snail's pace was not lost on me. I only popped back up on deck after dark to watch the necklace of ship and shore lights glisten around the harbor. I was wistful.

My phone rang at 0530 the next morning. "Most here. We go in today. You come here first. You get sun cream before you go out there—we are near the equator." He hung up, and I leapt out of bed. I could hear my mother's voice in his.

Given our proximity to the equator, and that I'd planned to stay on deck for the entirety of the eight-hour journey through the canal, I wore as little as I could, short of putting on a bathing suit. In denim cutoffs and a halter top, and with my hair pulled up and back in a pony tail, I raced down to eat, picked at my food, and stopped in the galley to grab the gloppy white stuff from Herr Most. He just grunted and shook his head at me as if I were his teenaged daughter borrowing his car. I headed out to the deck to watch the show. Many of the crew who were off duty had already crowded outside, and I slowed just a bit on seeing them. It didn't occur to me that I'd have company, and I felt like a dolt for thinking I'd have a private screening, and a little edgy at being this exposed to them like this all day long.

"*Morgen*," they said, one by one, and nodded. I felt like a new prisoner being paraded by the veteran inmates.

"*Morgen*," I said to each, making only the briefest eye contact.

Karl, my tablemate, stuck close to me. He had turned out to be a sweet man; we talked more at each meal, and I appreciated his quiet, thoughtful demeanor. He seemed much gentler and less rough around the edges than some of the others. His proximity felt protective. From the others, it would have felt threatening.

"The pilot will come on the ship soon to guide us into the first set of locks," he said.

I had to stand up on something to be able to peer over the side to see the small pilot boat sidle up to us, and the pilot and his mate board. They would assume control of the ship from Kapitän Beucking until we were safely through.

We approached the entrance of the first lock like a thick thread aiming for the narrow eye of a needle. The canal sat above sea level, so in order to pass we had to enter a concrete rectangular set of chambers to lift us up at the entrance and then lower us back down at the end. Each lock had its own name: Pedro Miguel, or Mira Flores, for example.

Just as we reached the entry to the locks, a coterie of navy-khaki-clad men carrying Herman Munster lunchboxes swarmed over the ship like ants on picnic watermelon. Like Lilliputians securing Gulliver, they had to tether the ship to the lock's "mules." These caboose-sized machines operated very much like train engines, but they were pulling ships. They flanked us and would slowly pull the ship into and later tug it out of place in the lock.

When the ship entered the lock, she sat low in the water, with the top of the walls on either side of us rising way above her main deck. Immense, solid metal gates closed and sealed behind us, and then water flooded in, like a tap filling a bathtub.

Slowly, slowly she rose, buoyed up by this miracle liquid that could raise the behemoth with the ease and grace of Mufasa hoisting Simba overhead. We could see the progress by watching the walls drop away until the main deck sat eye to eye with the top of the wall. The ship rose like Frankenstein on the slab, rising up out of the roof to attract lightning. Only when we reached this height, that of the canal water above sea level, could the mules tug us on our way, like a slow-motion slingshot, to the subsequent locks, and eventually out into the open waters to wend our way forty-eight miles toward the final set of locks that would gently set us back down to sea level and send us out on our way to the Pacific.

Curious about the canal and its operation, and desperate for conversation in a language I knew, I gathered the courage to approach and speak to one of the cadre of Panamanian workers on board. I picked an older, gentle looking man who sat and ate a mango and read a newspaper, taking a break while we passed through the lock.

"*Buenos días, señor ¿Como está*? I said.

He looked up; surprised to hear Spanish come out of my mouth.

"*Bien, bien señorita, y usted*?" He smiled kindly and offered me a piece of squishy ripe mango on the tip of his pocketknife, as if he and I were old friends. His dark chocolate skin was wrinkled, but his short hair was still mostly black. I could not fathom how he and his coworkers could wear long pants and long-sleeved dark shirts—his was buttoned up to the neck—in this cloudless, relentless heat, but he seemed cool. He squatted, agile, not resting on anything but his heels.

"*Oh, muchas gracias, sí*," I said as I took the offering, the sticky nectar dripping down my arm. "*Me llamo Diane.*"

I nibbled the tropical treat (we had little fresh fruit on board) and we chatted easily. I asked him about the canal, the surrounding area, and his work. He had been doing this since he was a teen; I guessed he was near sixty now. He thought I was Panamanian—the near-equatorial sun had toasted me—and hoped I'd have time to visit in Colón. Sadly, no, I said, not this time. So many people pass through here, and so few stay. He asked how I'd learned to speak Spanish so well, and guessed that I was fifteen. Maybe later in life I'd enjoy having my age underestimated, but now it made me feel like a child. He had to get back to work; break was over. But first he asked a question that I would hear over and over all summer, and asked myself often.

"*¿Entonces, qué haces aquí?* What are you doing here?"

"*Trabajo.* I'm working," was the simple answer. I'm not sure I even knew the complete answer, but it was more complex than I could convey in Spanish.

I'd wanted to get away from suburban New Jersey, with its quaint downtown and gray strip of uniform megastores on Routes One and Nine. I wanted to see the world. To have an adventure like many of my Middlebury friends had done, and I hadn't. I wanted some time and space between me and my old high school friends and my new college ones. Getting into and attending Middlebury had been a step up for me—a promotion in the social stratum. I looked so New Jersey compared to the corduroy-and-khaki-clad preppy kids with their seemingly perma-starched LaCoste shirt collars underneath Fair Isle sweaters. Maybe my tenure here, after feeling like a complete outsider on board in every way, would make me more of an insider at Middlebury.

Certainly I was here to make Daddy happy. He had seen this as a once in a lifetime opportunity that I should not pass up. So, good girl that I was, I didn't. But it didn't necessarily assure his approval. He would just move on to the next thing. The fact is, I had no idea what this trip would entail, or what I wanted from it, and wondered, just like everyone around me, why I was here. Like the ship, slung into the locks and tethered to the mules by thick cables, I was largely going along for the ride.

He offered up the newspaper he'd finished reading. I accepted, as thrilled with it as with the mango. Getting through that much print in Spanish would take many hours. I needed a way to fill all the hours alone in my cabin facing me on the other side of the Panama Canal.

"*Gracias, mucho gusto*," I said.

"*Nada, nada. Vaya con Dios*," he replied, and his broad smile lit up his face like a small child's.

I went reluctantly back inside to cool down and visit the head, rushing like a moviegoer loath to miss even a moment of a riveting film. Inside, another local man sat on an overturned bucket selling postcards, candies, and beer. Some of the crew surrounded him, perusing his wares. I wished I'd had money to buy some postcards to send home, but I didn't want to waste the time it would take to go up to my cabin to retrieve it.

I resumed my post at the bow, trying to avoid some of the more leering, lecherous-looking, dark-blue workers. I stuck closer to my own crew, and for the first time felt some safety near them.

Once the locks had lifted us up to the higher water level, they loosed us into the section of the canal that looked like a lake, with seemingly unfathomably deep water, into a real-life version of Disneyworld's Jungle Cruise.

The banks overflowed with lush, dense, exotic greenery. Ibises and spoonbills, egrets and herons, frigate birds and falcons swam nearby and flew overhead. Lighthouses and small thatched huts dotted the shore. I half expected some animatronic Mowgli to peek mechanically in and out of one. We were on a smooth river cruise, in no rush to get anywhere, it seemed. The reality was very different. We had slowed from an open sea speed of twenty-two knots to a crawl closer to five knots in the relatively more confined canal. The ship had appointments to keep at six ports in Australia and New Zealand, but for now, the tight banks and thick traffic kept us from rushing.

Other vessels kept us company and came into sharp focus as we neared and passed them going in the opposite direction, including several banana boats. I felt so ignorant not even knowing there was such a thing. Every day revealed how much I did not know.

A long oil tanker trailed us all the way. I imagined no matter how many times I'd make the trip through this man-made continental breach, it would never become routine or mundane. There was something new and fascinating around every bend.

The other vessels' crews hung over the railings, as we did, to take it all in. We waved at each other like gentry promenading on the QEII, but the more I saw of the others, the happier I was to be with mine. They were greasier and dirtier, in wife beater tees and stained, ripped pants that no one on their ships mended for them. I imagined they smelled unsavory.

Whenever they spotted me among the crowd, their looks changed to something more lascivious and their waves invited something darker than casual greeting. I did not see one other woman the entire day. As my relatively cleaner and clearly more disciplined crew stood steadfastly behind me, I shuddered to think how different my experience might have been on one of those other ships.

The passage took all day. I went in only to eat quickly, and maintained my post for most of the trip. My slack-jawed awe and amazement amused the others.

"Your first time?" asked Tim, with overtly sexual overtones. They all tittered.

"Yes, Tim, I'm a canal virgin," I answered, deciding not to cower at his kidding. More tittering.

"You will not be so amused the next time." I didn't reply. I didn't want to play.

———————

The canal narrowed from a lake to a wide river as we approached Panama City on the Pacific side. Those locks worked in reverse, letting us down easily. We approached and crossed under the vast bridge that made up part of the 19,000-mile Pan-American Highway, running from Prudhoe Bay in Alaska to Ushuaia, Argentina. Children swam in the water near the bank under the span and waved at us as we passed into the Pacific.

And right beside us, as if on cue from Herr Rose, the iridescent, fantastical flying fish appeared, leaping alongside the hull as if to escort us out into our seventeen-day trans-oceanic crossing. They lifted out of the spray at rhythmic intervals, like choreographed lords a-leaping, thin fin membranes spread open eighteen inches wide like wings. They alternated jumping out, gliding, and diving back in with their schoolmates, so the surface shone and quivered with their ballet. Airborne for what seemed like an extraordinarily long time—thirty or forty seconds—their dance welcomed us to this side and sent us on our way. Once through the canal, we'd have nothing more than a very few dots of islands to see for seventeen days. Nothing but blue. Sea, sky, and me.

———————

Maybe because I was aware that the Pacific was nearly twice the size of the Atlantic, the horizon seemed to loom larger and longer once we'd sailed under that bridge. This was it. The point of no return. I was leaving my country, my continent, and everything familiar behind. We were way out to sea. Up until then, there had been frequent and somewhat familiar touch points to anticipate—visible signs of progress. But once the canal was in the rear view mirror, the only thing I'd see before Sydney would be a brief glimpse of the Galapagos and Tahiti. The ship and its inhabitants would be my world, which in this almost incomprehensible vastness, suddenly shrunk, containing and confining me. Like an inmate, hands cuffed, I could hear the prison gates slam shut behind me.

A post-partum-like depression settled in after the excitement of the canal as I returned to my cabin. I was rank with sweat, having sweltered for nearly eight hours in the searing sun. The water had broken, the cord was cut, and I was soon to emerge into a new hemisphere. I made my way inside slowly, reluctantly, a little suspicious of what lie ahead. Every step of the nearly vertical internal staircase and down the blue hallway stretched out like the twists and turns in a distorted funhouse.

If my head hadn't hung down as I approached, I'd have seen it. But as it was I nearly stumbled over the bulk as I went to insert my key in the lock. My eight-track tape player, with a note from Herr Stuhlemmer, sat at my feet.

> *"Fraulein Meyer:*
> *Electrician Hoeldlmoser spliced the wires and attached*
> *the correct plug. Your machine should now work. We*
> *can replace the old plug when we return to New York*
> *if you desire."*

My music. I fumbled with the key as I cradled the cumbersome cube in my arms. Kicking off my flip-flops, I dropped to my knees, plugged the machine in, and slid it into the cubby in my nightstand. I grabbed a tape—one of only a few I'd brought on board—pushed the power button, and the "on" light smiled at me. Magic. No longer completely alone in my cabin, Genesis, Pink Floyd, Jethro Tull, and a few other familiar friends would keep me company.

The song came on.

Like Peter Gabriel, I too, wondered where, indeed, did my country lie? I may not have been homeless, but I was temporarily stateless.

The lyrics filled the air, filled my ears, and released something I'd been holding back for a week: tears. Even through the cheap tinny speakers, it was the sweetest sound I felt I'd ever heard. The melody mingled with my tears.

Selling England by the Pound. Genesis. I sang along to the endless loop of music that, as I dove deep into this ocean, offered some salvation. A thread I could grasp to maintain a tenuous connection to the people, places, and language I knew. To the life that felt comfortable and familiar, not so foreign in every way as this one.

It made me think of my high school friends Lauri and Sharon. We had listened to this record until we'd worn the grooves bare. Sharon and I were huge Genesis fans. We'd seen them on the *Wind and Wuthering* Tour at Madison Square Garden. Sharon drowned on her birthday in a bathtub during an epileptic fit. In such a small volume of water. And me here alone drowning, in a different way, surrounded by so much of it. Her six-year-old brother found her. She'd said that she didn't want to turn nineteen. She'd stopped taking her medication. I didn't go to the funeral because I was afraid to see my friend in a coffin. To see my friend buried. Being mid-semester at Middlebury was a convenient excuse that I felt guilty about forever. Lauri flew down to Texas where Sharon had recently moved and more recently died. I'd never forgive myself for not going. Now just Lauri and I were left from that triad.

I lay on the floor and listened to the album all the way through, over and over, the words washing away the grime of the day, the tears, and the fear. To this day, no other album has a more visceral effect on me.

When I showered for dinner, what I saw in the mirror horrified me. The sun cream that Herr Most had given me got no nearer my skin than the men hanging over the railings of the ships we'd passed. I was mahogany. I did not recognize myself, and understood why my friend Manuel thought I was Panamanian. I thanked my Greek heritage for the fact that I was not glowing radioactive red, and I was also very grateful that my mom could not see me now. She'd be furious. And Herr Most would surely be so in her stead.

At dinner the men's mood felt lighter than mine, lighter than it had been before the canal. The passage seemed to have loosened some unseen tension valve. A bottle of the ubiquitous Holsten beer sat at each place. Usually, the captain treated everyone to a beer on a crew member's birthday, but we found out about those early in the day. On the single sheet of onion paper announcing the ship news that greeted us at breakfast. I was unaware of any celebrations that day.

Ingo came over, amused by my surprise as I gaped at the beer, unsure if it were for me and if I ought to drink it. "*Prost!* Drink up!" he said. "Herr Kapitän Beucking is relieved to be though the canal. It can be tense for captains, you know; they are not in charge of their ship."

I nodded my head, even though I had no clue how ship captains felt.

"Now he is happy that we go full speed. So he treats everyone to a beer!"

I touched my fingertips to the droplets of condensed water dripping down the side of the dark brown bottle. Its logo, a black knight with a feathered helmet atop a rearing steed, armored with a red shield with a white "H" emblazoned on it, adhered to the glass. Red and white, just like the ship.

"You have tasted Holsten before?" He asked.

"No, never."

"You have tasted beer before, Fraulein Meyer?" He asked, his grin impish.

"Yes, Ingo, I have tasted beer before."

"Drink then! *Prost!*" he said, and he released one of his deep resonant belly laughs.

"Prost!" We clinked bottles.

Oh, that beer! It was as icy as the day was sultry, manna from Holsten Brauerei, AG. So different from the Miller my father drank. The few clandestine sips I'd had of that had kept me away from beer until we all started drinking whatever cheap brew someone's older brother snuck through the small basement windows in high school. At Middlebury we sipped watered-down piss that we ordered by the pitcher for $3 at the Rosebud Café, or dispensed from frat party kegs. I never knew what real beer tasted like before that first Holsten, and it delighted my palate and set me up for disappointment for the next two years at college. By the time I'd finished it, my shoulders had un-hunched, and I began to think the next seventeen days might not be so bad.

On his way out of the mess hall, Tim stopped and put a hand on my table.

"*Abend*, Diane."

I looked up, surprised. He and most of the crew had only greeted me casually and in passing. I didn't realize he even knew my first name.

"*Abend*," I said.

Long, lean, and lanky, Tim was a skeleton hanging in an anatomy class thinly cloaked in skin. His coveralls were always too baggy and too short. He corralled his shoulder-length strands of dirty blond hair with a dingy red bandana tied across his forehead. His downward-focused, doleful eyes made me think of it more like a crown of thorns. He had an air of Christ about him: the suffering, not the sacrifice. The puncture marks on his limbs were not from spikes but from heroin needles. This ship was his treatment center. He was here to rehab. His spider-leg fingers spoke of the work he did, the lowest kind on board. Paint, grime, and calluses competed for space, all equally reluctant to give up or go away. He was hard. Hardened. Weathered, like the ship that he and his mates painted on every single round trip to ameliorate the impact of the punishing salt air. But he was not so lucky. No amount of paint could erase the damage that his life had inflicted on his face.

He was British, but his tenure on board had teased the cockney accent out of him and created some sinister hybrid, an unnatural GMO sound, neither here nor there, like him. Did his wife and daughter miss him? If so, he did not seem to reciprocate. Tim was Cool Hand Luke. The hard-living, soft-spoken leader of the pack. He saw all and answered to none.

"We're having a small party in our cabin tonight. We'll play chess and talk. You can join us," he said. Not so much of an invitation as an edict.

Still feeling badly that I'd demurred after Kino Night, and before I thought much about it, I said, "Thanks, where's your cabin?" I really hoped to socialize more and spend less time alone on the long crossing. He gave me the cabin number and told me to come down any time after dinner. But my shoulders re-hunched after accepting. I had no idea what to expect, or what I might have gotten myself into. I dreaded wallowing my way across the Pacific in isolation, but the alternative scared me, too. I dressed down so as not to rile anyone up.

The mood in the cabin was festive, the lights, muted. The captain's relief must have trickled down to the common folk who intended to revel in it.

The difference between my lush digs and this hovel filled me with guilt, the way the hoppy smell of spilt beer filled my lungs. The entire cabin would fit twice in mine. I wondered if they knew this. Wondered if they hated me for it. The twin beds, narrower than mine, took up much of the uncarpeted floor space. The only light, day or night, since inside cabins had no portholes, came from a dim overhead fixture and reading lamps affixed to the wall above each bed, making the scene look like it had been shot with a cheese-cloth-covered camera.

Ten or so people crowded into the narrow space, using every available surface: the beds, the small built in desk, and the two desk chairs. Elton John sang in my ear: *"They're packed pretty tight in here tonight,"* as Tim introduced me around. I was relieved to see Claudia; at least I wasn't the only female in the space.

I recognized Chris, a cute friend of Tim's, who reminded me of the guys at college. He was tall, fit, and carefully groomed. He smiled easily, and laughed guilelessly. His *doppelgänger*, Roland, was shorter, slighter, and very reserved. He always looked caved in on himself. He could have joined any hair band with his shoulder length brown curly locks. His eyes, always at half-mast like those on *The Great Gatsby* billboard, saw and judged everything, silently. Chris was a happy, wagging, yappy puppy; Roland, a distant, inscrutable Cheshire Cat.

Tim handed me my second beer and sat me down. I watched Roland beat one of the others in only a few moves at a chessboard they'd set up between the beds. His hands barely moved, his face even less.

"I'll play you," said Tim when they'd finished.

"Oh, no, I. . . ." but again, it was more of a statement than a question.

I played, with all eyes on me, especially Roland's. He winced almost imperceptibly at every move I made. Tim made quick work of me, and Chris said,

"Let me play her! Let me play!"

Someone handed me another beer—that was three, and not having had a sip since leaving school, I was feeling just fine. Maybe too fine. I did a little better against Chris, but still lost.

"You go, Rollie, you play her," Chris said to Roland, as if I were a chew toy he wanted to share. Roland looked down from the corner where he stood, and shook his head, like Rick denying a patron access to his Café Americain in *Casablanca*. Chris drooped as if he'd been whacked on the bottom by a rolled-up newspaper.

I moved over to sit near Claudia.

"Hi," I said.

"Hello."

"How are you?" I asked, feeling awkward and strained, but very much wanting to make friends with the only other girl I felt I had even a small chance to connect with. I never saw Ana on board, except sometimes by the small pool in a smaller string bikini.

"Good," she said, after a gulp of beer. "You?"

I felt certain that out of everyone on board, it was likely that she felt most wonder about why I would put myself voluntarily into a situation that she pretty clearly did not relish. She was Austrian, as was her husband, Bruno. I supposed that if I'd understood German I'd have detected a different accent, but my ear was blind to it.

"Where is Bruno?" I asked, and then regretted the too-personal intrusion.

"Oh, he sleeps. He wakes very early to make breakfast, and. . . ." She paused and took another swallow. "And he does not like so much, um, a party," she finished, and looked down at her beer.

There was something sad about her, and something she wasn't telling me, but I sensed that it wasn't because she couldn't figure out how to say it in English. I didn't want to pry, so I shifted the focus to me.

"Yeah, I wasn't sure if I should come down. It's a little weird for me on board."

She looked up at me.

"*Ya, ya,* for me too sometimes."

"But at least you speak German," I said.

She smiled. "*Ya, ya,* it is better."

It was late. I was swimming in a pool of that strong beer. It seemed like a good time to go. The close cabin felt like the inside of one of the containers we carried on board.

"Claudia, it was really nice talking to you. Thank you."

She looked at me as if to say, *thank you for what?* She had no idea how much it meant to me to feel even a small hope that I'd have a friend on board. It would be a real lifeline in this sea of men.

"I'd better go. I have to be up early, to see what Herr Most wants me to do." I wanted her to know that I was not sleeping late and lolling about all day. It occurred to me that they had no idea what I was doing on board, in both the narrowest and broadest senses.

"*Ya,* me too," but she made no motion to stand. I saw her look out of the corner of her eye at Tim, who returned the glance.

"*Gute Nacht,*" I told the group, raising minor objections from a few.

"Not time for bed yet," they said.

"*Danke,*" I said, and slipped out of the door.

―――――――――

The wall phone sounded especially loud. I looked at my little clock: 0530, for the second day in a row. No way. My head hurt. My mouth stuck closed.

"Most here. You come up to the galley now. *Schnell.*"

Shit. He's kidding, right? Never a "please." I must be in trouble for going to the party. Normally I'd report to him at 0800. What could he need now? He hung up before my tongue loosened.

I stumbled my way into my clothes and grumbled my way out the door. He took one look at me and burst out laughing.

"You make drink too much last night, *ya*?" So Herr Most shared with my mother both a passion for sunscreen, and the uncanny ability to know everything that went on anywhere. Was it that obvious?

I just looked down, chastened, and not in the mood for this chiding.

"I talk quietly then, *ya*? When you go with the crew you make them buy the beer, *ya*? You don't pay." He handed me a glass of water.

This tack took me by surprise. He was being nice.

"*Danke.*"

"But Fraulein Meyer, where did you put sun cream? Not on you, I see," he said, lifting and rotating my milk-chocolate arm as if examining fish in a market.

"Um, I. . . ." I started. *Why the fuck am I here at oh five forty hours? Surely not to talk about sunscreen. I'm tired. I have a giant headache.* I did NOT say this.

"You use it next time, *ya*?" he said as he dropped my arm and handed me two aspirin to go with the water.

"We have not so much mending now."

Because of your thorough sewing skills, I knew he'd never say.

"So now that we are past the canal, you will have new jobs. You will report to the bridge every morning at 0600 hours to make clean before Herr Kapitän reports. I will show you this morning. Also you will help me with Kanteen. And you will mop the hallways and stairs inside, *ya*? And clean the walls. Claudia will help."

No way. I'd actually be swabbing the decks? The cliché materialized. So the kid gloves were off. But I could hardly complain. It was all inside the air-conditioned superstructure, and it was still a fraction of what everyone else on board did.

"Ya, ya," I said. But at 0600? Couldn't the bridge wait until, say, 1500, to get clean?

"After bridge, you eat breakfast. Then come to me. We go up now, to make clean."

The bridge was narrower than it was long, an almost

semicircular room atop the superstructure. It afforded a spectacular, panoramic, nearly 270-degree view through large, invisibly seamed picture windows. Looking out toward the bow where I'd spent most of the previous day gave an awe-inspiring perspective on the containers we carried. And I could only see those that were above deck. For nearly each one I could see, I knew there was also one below. It humbled me with respect for this scarlet lady. The bow bobbed up and down from this vantage point, gently, and in time with the silent music of some invisible sea muse.

Sturdy, dimpled, plastic flooring steadied our steps. The burled counter below the glass panes clicked and flashed with an array of incomprehensible gauges, levers, and lights. Behind a railing, and us, a shiny silver wheel the size of an extra large pizza was mounted on a dais. Even though it didn't have wooden spokes, like the ones in old pirate films, I could see Captain Hook and Smee at the wheel. Everything else about the scene screamed movie set. It was a little surreal to be standing where I could watch the bow dip up and down like a novelty bird at the rim of a cocktail glass.

Already dazed by the hour and the hangover, this just struck me dumb. Herr Most had to nudge me out of my reverie.

"*Morgen*," he nodded deferentially to Second Officer Betz, who was finishing up his overnight shift.

"*Morgen*," he nodded back.

Herr Most led me around the bridge with a collection of cleaning supplies to show me how to make it shine in the most unobtrusive way possible. He forbade me from getting in the way of the duty officer.

I became intimate with and made that bridge glisten almost every morning for the balance of the trip, unless we were in port. I did not mind the work. It was boring, but satisfying, as I could easily see when my efforts turned something from dusty or dull to shiny. I sensed that Herr Most wanted the officers to see me busy. I wondered who had done this before. Claudia? Whatever phantom did the heavy inside cleaning?

The officers' eyes on me while I worked made me feel awkward. The bridge was quiet at that hour; the ship on autopilot, I imagined. The duty shuffled papers and checked the instruments periodically. The silence magnified every noise I made. Officer Betz rarely said more than "*Morgen.*" Officer Rose would make small talk, but whoever was there was at the end of an overnight shift, and generally in no mood to chat.

One morning, Herr Kapitän came up early. My cabin sat right next to his suite, and sometimes I could hear him at night. I wondered how he became captain, what he thought about his job, what he worried about most, and whom he missed. If he reciprocated the curiosity, he didn't let on. He nodded politely and curtly whenever we passed. He was a German Clint Eastwood, tall, tight, and rangy, with weathered leather skin. His blue eyes were permanently narrowed to a squint behind the rectangular, gunmetal-gray, wire frames of his glasses, and his jaw clenched tight all the time. I never saw him smile, and I never saw him wear anything except his dress uniform.

My stomach tightened to match his tight temporomaniblular joint. Betz jumped up and stood tall. "Herr Kapitän," he said.

"Betz," he nodded, and then noticed me, in sloppy shorts and T-shirt, hair pulled back and hands covered in yellow rubber gloves. I stopped wiping the glass that covered the gauges.

"Fraulein Meyer," he nodded again.

"*Guten Morgen*, Herr Kapitän," I said, not knowing if I should bow or curtsey or salute. I fumbled to remove the glove from my right hand, but by then he'd turned away and picked up some paperwork to study. Relieved, I finished up quickly and scurried off the bridge, not quite sure why I felt like I ought to have my tail between my legs.

The next morning, I squinted at my reflection in the small medicine cabinet mirror, its edges rounded and rimmed in brushed chrome, to see cracks. On my forehead. Not in the mirror. It looked as if those Lilliputians from the canal had returned and plastered my brow in tiny taupe tiles—or maybe rectangles of thin tissue paper, with flesh colored-grout in between. They had not done a very good job, though. As I peered more closely, I could see that the tiles were peeling up at the corners. I put my hand up to touch my brow, I guess to be sure what I was seeing was real, and several pieces flaked off like phyllo dough from the edges of baklava. I'd literally baked myself in the Panama sun. No amount of nagging from a parent or other adult (say, the Chief Steward) could have had the impact of watching the better part of my face flake off into the porcelain basin below. I didn't even need water for the sloughing. Gentle exfoliation with my washcloth filled the sink with the miniature brown paint chips falling from my face. Well, a clean slate, and a strong reminder to use the sun cream.

––––––––––

My favorite duty by far was assisting with Kanteen, the small on-board supply shop that felt like a Playskool store. Herr Most and I stood in a closet-sized storage room, lined floor-to-ceiling with inventory, behind a closed half-door featuring a protruding lip that formed a small counter. A metal hook secured the top half of the door to the outside wall so it didn't constantly slam open and closed with the ship's motion.

The Kanteen resembled a cross between a Little League concession stand and a military PX. We sold cigarettes and condoms, shampoo and stamps, beer and Barbasol. I had wondered where the crew got their brew, and how they stored enough toiletries in their small quarters for an entire trip, but now I understood. They didn't always have time to run ashore when we docked, since we were never in port for very long, and not everyone had the time off. So the ship made this service available, and made a little profit, too.

We stood together, Herr Most and I, shoulder-to-shoulder, since he was only marginally taller than I. We took each man's order and Deutsche Marks, and handed him goods and change. Like playing "store" in the backyard with Suzanne when we were kids. Selling leaves and twigs to faux customers in exchange for pebbles.

He barely said a word to anyone; just nodded and provided the perfect foil to my Chatty Cathy. It afforded me the opportunity to have more social interaction, but from a safe distance behind a barrier. I got to see some of the crew with whom I rarely crossed paths, and I studied each one, as though I were cataloging them or keeping inventory, like I did in the shop. Most of them ignored me, but some smiled and engaged in conversation. Alois and Chris always seemed to need something, and tended to linger over their orders. Although they were making it fairly obvious, I tried to ignore the fact that they'd both started to compete for my attention. The former in a goofy, awkward way—always poorly timed and just a little off. The latter like an overzealous, panting dog.

"Fraulein Meyer!" said Alois. "*Wie geht es*? So nice to see your lovely face here to serve us."

Herr Most looked up and glared at him and frowned. Alois stopped grinning. "Herr Ganser, what do you need tonight?" he asked in English, blocking any further attempts at small talk.

"*Abend*, Fraulein! How are you? We have Kino night tomorrow! Do you go? It is an American movie, I think you will like it!" said Chris.

Herr Most cleared his throat. All business, this man. Move along, said his body language. But I sensed, too, that in some way he was part of the barrier that kept me safe.

I'd never, even in the Middlebury frat houses, seen men consume so much beer. We sold more of it at Kanteen than anything else. It cost less than soda. We could never keep enough in stock. Whence, I wondered, came this endless wellspring of ale? And then one day Herr Most walked in while I was setting the table for the officers' lunch.

"Come."

I looked down at the silverware in my hand, unsure if I should leave it or finish, and again he said, "Come. Zip, zip."

I dropped the forks and followed behind him. I never knew, when summoned, what to expect, and always felt just a little afraid that I was in trouble, even though I'd never done anything to warrant admonishment (except for my glaring failure to use sunscreen). "*Schnell*," he said. Where was the fire? Why the urgency? But I'd begun to realize that Herr Most's default mode was "urgent." It's probably what made him good at his job, and kept him so thin. His metabolism had to rev high to keep up with him.

"We need to restock the Holsten. You will help."

Ah, he could have said that to begin with.

We descended further into the depths of the ship's hold than I'd ever been before, wandering through a red intestinal maze of pipes and other tangled protuberances, over metal floors embossed with interlocking petals for traction. Herr Most stopped abruptly in front of a large padlocked door and unlocked it with a key from the ring that he, the gatekeeper, carried at all times in his pocket. A welcome rush of chilled air counteracted the gray steaminess that clutched us.

He turned to me and motioned. "Come." I stepped into a pitch-black cool cave. A container of different sorts below deck. I expected to trip over stalagmites and bump my head into stalactites, dodging furry black bats all the while. Until he flipped the switch and revealed another movie set: a warehouse scene straight out of *On the Waterfront*, but it wasn't ship's cargo. Scarlet cases of Holsten, emblazoned with that familiar knight, sat stacked from floor to ceiling, so deep that I couldn't see the walls in a space the size of a small elementary school gym.

"We have six hundred sixty-three cases here." He said. My Middlebury friends would have had to pinch themselves, as I was tempted to do, to confirm they weren't dreaming. And then they'd proceed to get so drunk that they'd pass out right here in the giant refrigerator.

I struggled to do the mental math. I gave up the fight, and just said, "That's a lot of beer."

He chuckled and retrieved a large hand truck that stood strapped to the wall. "You load them here." We stacked as many as we could, and took the elevator up to the galley level, where we stored some, and stowed the rest in the Kanteen.

He clearly had not needed my help. My brawn hardly facilitated the task. He just wanted to show me the skyscrapers of suds, but could not allow himself to simply say: "Hey, come on down, I want to show you the ridiculous amount of beer we have in an enormous cold storage locker." He would not tip his hand by showing that much enthusiasm, nor let me know that he cared enough to amuse me. A sweet heart beat somewhere beneath that icy veneer, but he took care not to let the shell melt.

I now knew that no matter what else happened for the balance of the trip, we would not run out of beer.

———————

At 1500 hours every day, all activity ceased and all hands streamed into their assigned mess hall, or in my case, the galley. Every single day. I had never heard of this custom before, and was unsure if it were a German tradition, a nautical tradition, or neither of these. It had food and hot beverages in common with British tea time, but it occurred earlier in the afternoon, included coffee but not tea—although we drank that with every meal—and had the odd (to me) adjunct of the cigarettes. Nearly everyone smoked. I could not get over that. Even back then, we knew the dangers of smoking; very few of my friends smoked. And I couldn't understand how the ship—especially a meticulously spotless and disciplined one—allowed a practice that was both so dirty and potentially dangerous. The irony of being surrounded by all this water that would be of exactly no help in case of fire was striking. Perhaps the vice was too ingrained in the culture for them to even consider prohibition; perhaps the captain smoked, and it was, after all, his ship; but whatever the reason, smokers prevailed.

Herr Most would say, "Schmoke time!" when I walked in, as if I'd be surprised. Like a Pavlovian dog, I'd follow the scent of fresh-baked goods there on cue, salivating all the way.

He would have already laid out the goods for the officers, but always saved a stash of whatever Ingo and Bruno had baked that day. No Toaster Strudels, these, the waft would usually clue me in to what awaited. Cinnamon-scented, gooey apple or raisin strudel, crisp and flaky on the outside, soft and doughy and warm inside, drizzled with sweet white icing. Or cookies—chocolate chip or sugar, the latter frosted, too, with steam still rising from them. Or layered Napoleon-like concoctions with thick rich cream hiding in the layers. I'd sneak in, almost furtively, as if Herr Most and I were keeping some secret. "Schmoke time!" he'd say, and hand me a plate of pleasure and a carafe of coffee, and then say, "You bring the plate back, *ya?*" as if this were the first time, or as if I'd failed to do so every single day so far.

"*Ya, ya, danke.*" And I'd slip out, like a squirrel with an acorn, or more like an addict with a fix, to go back to the peace and privacy of my cabin, where I'd put on some music and stuff the confection into my hungry cheeks, unable to mainline it directly into my veins. I never knew why he let me come to the galley and take the treats back to eat by myself, rather than just making me join the crew in their mess, but I was eternally grateful. Or perhaps I shouldn't have been. Maybe if I'd had to eat in front of everyone else, I'd have exercised some self-restraint. Or exercised my body more. As it was, no watching eyes prevented me from indulging.

These sugar-filled respites provided me some shelter from the storm of conflicting emotion that buffeted me daily. Forget the Freshman Fifteen. Even with a nineteen-year-old metabolism and a moderate level of physical work, I gained the Nautical Nine in no time.

In seventh grade Doug Bauer, the first boy I kissed in Tamaques Park on the way home from school, told me I had fat thighs. We'd sat down on my front steps and, in cutoffs, gravity had flattened them out on the concrete. Self-conscious ever since then, I could not go back to school stuffed with pastry cream. But I also struggled with giving up one of the few comforts on board.

"I need to stop eating those pastries!" I noted in my journal. I started doing half-hearted push-ups and sit-ups in my cabin. These were no match for the sugar, flour, and butter. I could no more break this habit than an addict could quit heroin cold turkey. There were few pleasures on board; I was loath to give this one up. The sweet on my taste buds triggered serotonin that was sorely lacking otherwise.

I would slink back each afternoon, sated in a sugar stupor, leave the empty dish and carafe on the stainless-steel counter in the galley, and go back to my room, ashamed, to sleep it off.

The crew felt more comfortable with me on board; this was a blessing and a curse. I was very mindful of what I wore and how I behaved. I really enjoyed getting to know my shipmates and learning more about their lives on and off the ship, but I worked hard not to give a false impression of what I wanted from them, or what they might expect from me.

There were two kinds of party: events that the ship sponsored or sanctioned, and the unofficial, informal beer bashes in someone's cabin. The former, like Kino Night or the "Western" BBQ, had themes and specific time frames, or might celebrate some event (like crossing the Equator). They seemed meant to keep up morale, provide much-needed distraction from work and the long crossing, and encourage camaraderie. High- and low-caste members mingled for these: any officer not on duty might show up, along with the lowest deck hands. I rarely saw the captain or Herr Most, and if they did show up it was almost ceremonial: they'd make a brief appearance and then evaporate. Also, I never, ever saw the Chinese laundryman and worried about how lonely he must be. He seemed to spend all his time below deck, like a mole. I never saw him, even at meals.

The beer and harder alcohol blurred the lines that delineated rank, and the more the men drank, the louder and looser they got. They also became bolder with me. Reticent during the tea-totaling day, they'd come and sit close to me, in casual clothes, sweaty from the heat and humidity, and reeking of whatever they were drinking. They were little boys approaching a neighborhood dog that had initially scared them, so they'd kept their distance. Once they realized that I was domesticated and tame, they came ever nearer.

The conversation was always the same: "Fraulein Meyer! Why would a young American girl want to work on a container ship with a bunch of German men? So what do you think of our ship?" Sometimes they'd ask about my life at home, but more often than not, they would fire questions and spit Schnapps-flavored saliva at me, and not really wait for answers; just sway a bit and offer me another beer and a cigarette, no matter how many times I said I didn't smoke. The more animated they grew, the smaller I became, folding my arms, crossing my legs, and trying to disappear into myself. I'd redirect the conversation by asking about them, careful not to be too intrusive.

But with each sip of Holsten courage, they would become more aggressive or more critical of American politics, sports, or people, and wait for me to either acquiesce easily or defend vigorously. I did neither, and tired of the game quickly.

"Jimmy Carter? A peanut farmer president, so silly!" They'd say. I just smiled as broadly as Carter himself, and said nothing. Frankly, at that time, I had more interest in academia and the arts than politics, so I had no retort, and even less in the way of interest in engaging with them. They had no interest in Socratic dialogue. They just taunted me to amuse themselves.

I stuck strictly to a three-beer maximum. No matter how many they offered me (Herr Most continued to insist that I not buy my own), I was determined neither to take advantage of their generosity nor to water down my defenses.

I would just excuse myself when things got raucous and one of my reliable defenders was not there to rescue me, often to objections that might include someone grabbing my arm or putting theirs around my waist. Especially if some of the shadier characters I didn't know well, like the hulking machinist, were present. I could usually count on curly-haired Karl or Herr Rose to come to my rescue with a joke or some other gentle intervention to deflect attention if they saw me in trouble. In their absence I'd twist away and slip back up to my cabin, locking the door behind me.

Often at this point my phone would ring: "You come back to the party!" the unidentified caller would say. "We miss you!" I could hear noise and unfriendly laughter in the background; a bunch of middle-school boys making prank calls.

"No, I'm pretty tired."

"We can wake you up!" Then I'd just hang up.

The beige phone started ringing more and more often in the middle of the night, waking me up in confusion every time. I couldn't imagine who would call me. During the day, Herr Most might have a job for me, the radio officer might deliver some administrative information, or Alois or Chris might invite me to a gathering. But late at night, it was usually drunken sailors dialing, and my muscles would constrict as tightly as that beige, coiled cord. I wanted to leave it off the hook, but the constant tone that created was worse. And I still saw it as a way to reach out if I needed help . . . but to whom?

The informal parties, usually in Tim's cabin, had the feel of a private back room in a club. These gatherings were by invitation only, and the officers never attended, except for Alois. He, like I, inhabited that unusual ether between the poles, and could slip easily between them, morphing now into one of the boys when it suited him, and then into a proper officer when protocol called for that. I wondered if the officers had parties of their own. I imagined them playing poker and smoking cigars, sipping whiskey, on a background of Day-Glo oil paint in somebody's black velvet suite.

Soundtracks that I recognized from home scored these low-key nights: The Rolling Stones, Led Zeppelin, The Who. For as much German pride as they boasted, I'd never heard German tunes at these parties. We would just talk, in a group or one-on-one, about ship happenings, politics, and culture. And we played chess. Claudia, her skin so light as to be almost translucent, often came to these speakeasies, but never to the deck parties unless accompanied by Bruno, and then she'd sit quietly in his shadow, a cigarette in one hand, and a Holsten in the other. I never saw Ana anywhere but at meals or stretched out by the side of the pool in her tiny string thong bikini, her straight black hair adding a modicum of modesty as it draped down over her shoulders and nearly down to her waist.

While I spoke mostly to Alois, Tim, and Chris, who spoke the best English—the others looked on in quizzical curiosity and occasionally asked a question, or for a translation to German—it was Roland who interested me the most, exactly because he seemed completely disinterested in me. He would hang back in the corner like a vampire bat, rarely saying a word, observing the group through half-closed eyes. He only engaged to play, and usually win, a game of chess now and then.

"Does Roland hate me, or what?" I asked Tim one night. He looked up and over at him in the corner. He shrugged.

"No, no. He had some trouble on another ship with a girl. He thinks women are trouble."

Okay, fair enough. I think these men could be trouble, too.

"Go talk to Claudia," he said. "She is upset about something. Maybe she needs a female shoulder to cry on." It could have sounded caring, but it didn't.

I sat down next to her and ventured, "Hi."

"Hello," she said, tears making rivulets across her fair skin.

"Hey, are you okay?" I asked, and she shook her head. She mumbled something about a fight with Bruno. Something to do with herring, I thought she said, but I had a tough time understanding her between her limited English, thick accent, and soft sobs.

"Let's go out on deck where we can talk," I said. *And avoid prying eyes*, I wanted to add.

She nodded, and I put my arm around her waist. We'd been drinking Black Russians, and she'd outpaced me by at least two to one. I looked up at Tim, whose hung head shook gently as he made eye contact with me, and then looked away at Roland. Fuck them. Girl Power.

Claudia and I had been working together to mop the halls lately, and Herr Most told us to clean all the walls while we were at it. She reported to him as well. Wax on, wax off. . . . I felt like Daniel-san. The crew could not help but snigger and comment when they passed the Cinderella twins with our mops and rags. It seemed to give them some great pleasure to see us both dripping with sudsy water and sweat as we swabbed the blue linoleum floors and beige plastic walls. We grew grumpy in the internal heat. She would quickly retort with German words I did not understand but was pretty sure were the ones I'd use if I had the nerve. We'd grown a bit closer, if only to close ranks and strengthen our defenses.

"What's going on, Claudia?"

"He is horrible to me. He gives me no freedom. He is hard to me. He gets so mad even if I just talk to another man. Who else I can talk to? There is no one else. Just because I talk to them it does not mean I sleep with them. I only sleep with Tim once, anyway. Do not get married. It is no freedom. Keep your freedom." She full on ugly cried.

I had my arm around her as we sat on the metal deck floor, knees pulled in tight to our chests and heads bent to avoid the rain that was now pelting us. The black ocean below, indiscernible from the sky in the absence of moonlight, was an inkwell so deep that Shakespeare could have written all 38 plays and 154 sonnets with it, still making nary a dent. Wait, what? She slept with Tim? Shit. Like it was no big deal. Wow.

I said, "It's okay. It's okay. You have me. You can talk to me." I so desperately wanted her to need me as a friend, to see me as an ally, so I could see her as one. But even as I said those words, I knew that in the morning she would pretend that nothing had happened, and we would resume our relationship as companion scullery maids and nothing more. I hoped she did not remember telling me about Tim. I tried to forget.

———————

Certain things helped to mark the passage of time and distance since, on most days, there was literally nothing to see in any direction except the ocean and the sky. As constant as they were, they changed constantly. Their blues spanned the spectrum from Arctic to indigo. Sometimes the crests were so calm that they looked literally like little waves of nautical greeting lapping at us, barely strong enough to peak. Other times, the forceful, churlish chop churned this way and that, slapping at us, angry that we'd invaded the aquatic space with our crimson metal bow.

The sun played with the water. Sometimes it sprinkled it with silver fairy dust that landed but then danced on the waves. Sometimes it laid down a shine so blinding that it looked like ice thick enough to skate on. Sometimes it hid behind clouds turning everything a murky, monotone gunmetal. The sky could be fickle, too. It would show a painter's palette of velvety blues, or an ominous swirl of charcoal, threatening rain. But unlike on land, where the weather moves over you, she just moved us through the weather, so that the calm cerulean always came back on cue. We needed other landmarks in a place without land. It could feel like Las Vegas on board, even with the brilliant sun and moon risings and settings. I could easily forget what day it was without Kino Night or Kanteen to remind me.

Since we were travelling west, we turned the clock back an hour every other day to keep pace with the time zones we crossed. At home, we college students struggled to keep up with the simple semiannual hour shifts. This accelerated time travel belied what felt like the glacial pace of the physical travel. Time seemed to lose its relevance—to feel more like an artificial human construct when we toyed with it thus.

But although the clock shifted as constantly as the decks below our feet, the menu anchored us to our place in the week. They printed it up and left it out for us to consult, but they need not have bothered. After a couple of weeks I memorized it, and it never varied unless there was a special event meal. I knew what day of the week it was by what Ingo cooked and Claudia placed down in front of me each day. Schnitzel (also, coincidentally our miniature schnauzer's name; I missed her so much every time they served it), meatballs in brown gravy, and bratwurst—each, almost always accompanied by spaetzle—had their assigned day. "If it's Tuesday, it must be . . . Potato Pancake Day." The cooks adhered to this menu so rigidly that when we crossed the international dateline on the way to Australia we just skipped from Monday to Wednesday meals, and on the way back, we had two cuisines *a la* Monday.

Once a week, Ingo proudly presented me with his version of steak tartare: raw ground beef, an uncooked egg in the divot Ingo had pressed into it with his meaty thumbs. He would put it down in front of me with a broad smile: "Raw meat! It's good for you, *Fraulein*! It makes you strong like

Ingo!" Oh, the E. coli I dodged. I did not relish this bloody meal, but ate it every time, afraid to offend.

Freshly baked bread and sweet butter always sat alongside bottomless carafes of green tea. The lunchtime table always featured a platter of gelatinous cold cuts in various shades of pink, studded with things I did not recognize. We might have fruit or vegetables on occasion, and of course we could count on the schmoke-time sweets, but even these, I realized, followed the strict pattern of assigned days.

On June thirteenth, just short of two weeks after departure, Herr Rose pointed out one of the Galapagos Islands while I was cleaning the bridge.

"And we will see Tahiti maybe in six or seven days," he said.

"It's beautiful!" I gazed out over the deck, over the gently bobbing containers, as I bobbled along in my own, toward a more vivid horizon under the brightest sun I'd ever seen. Even though we were only two days out of the canal, I knew we had a lot of ocean to cross, and it was somehow reassuring to see land again, however far away.

"*Ya*, and we will pass the equator today," he said. Such nonchalance for something so momentous. We would be at the exact spot where the hemispheres met. It mesmerized me, and I longed to know the exact moment of the crossing, but everyone else was completely blasé about it. Been there, done that. Even though I was an equator virgin, no one took much notice of my transition to womanhood. No fanfare. No celebration. I was grateful that Herr Rose had even thought to mention it. Herr Most also mentioned it to me matter-of-factly later, and watched, bemused, as I peppered him with questions about exactly when and how I'd know. *Such a silly American girl*, he must have thought. I watched for the water in the sinks and toilets to reverse and drain counterclockwise, as it does the moment you enter the Southern Hemisphere.

The next morning, though, Alois stood waiting outside my door at 0600. He startled me as I stepped out, groggy, to head up to the bridge, and I sensed this pleased him. But he also beamed with a different, preternatural self-satisfaction that I struggled to comprehend.

He had something in his hands, something the size of a small poster. His grin widened as he turned it over to reveal a most intricately and elaborately illustrated and lettered design. The god Neptune sat astride the waves, trident in hand, presiding over the proclamation below him. I could not read the inscription because it was all in German, but could make out my name, Diane Lori Meyer, in big, black calligraphy letters, and what must have been June in German, 14, 1979.

"It is an important event when a sailor passes the equator for the first time," he said. Wait, I'm a sailor? He said I'm a sailor! He continued, "It reads: From Neptune, the god of the diligent and the streams, who rules over the high seas at all times, with a powerful eye. . . ." Neptune was long-winded, but proudly and boldly recognized me as "a full-fledged equator-passer." The word's length and stature in German, *Aquatoruberquerung*, made it look weighty and impressive.

I tried hard, but could not stop the tears that welled up in my eyes. The gesture was as stunning as the certificate. Wanting to, but unable to hold back, I hugged him tightly.

"Oh, Alois, it is so beautiful! That's amazing! Thank you so much! I cannot wait to frame it and hang it at home!"

He smiled. I know he'd hoped for a big reaction. He was the least reserved of the reserved Germans, and he seemed genuinely pleased. He'd certainly earned it.

"It is customary on a ship to commemorate each sailor's first time. I had to convince Herr Kapitän (who had signed it with an elaborate flourish) to allow me to give it to you.

He thought I only wanted to do it to make you want to sleep with me." He laughed awkwardly.

My eyes, glued to the document, rose to meet his. His smile faded ever so slightly.

"That was really nice of both of you. Thank you. I'd better put it in my cabin (which, I was acutely aware at that moment, was formerly his cabin) to keep it safe, and get up to the bridge. I don't want to keep them waiting. Herr Most would be furious if he got word that I was late."

"*Ya, ya.*" He nodded nervously. "Well, *Guten Morgen,*" and he handed me the certificate and turned to go.

"Alois," I said, "*danke,*" and went inside to stow it in the closet between some shirts.

That old sea dog, Herr Kapitän Beucking. He was not only very wise, but he was looking out for me. It didn't matter whether it was because he actually cared, or just didn't want trouble on his ship. He had my back.

———

The ship's alarm bell sounded loudly early one evening, just as we were finishing up dinner. I'd never heard it before, so it scared me. Which was, I realized, exactly what it was meant to do. Stir you to swift motion.

"Come," Ingo said, grabbing me with his beefy hand, wiping the other on his white apron. "Lifeboat drill. You are starboard, *ya?*"

I nodded. None of his usual kidding around here. He was serious. I followed him to the starboard side, where half the crew mustered; the others had the port boat. All I could think of, foolishly, was *all hands on deck.*

From afar, we'd have all looked the same, lined up, covered in bright-orange, puffer life vests. But as always, I stuck out sorely in the crowd. The gathering highlighted how different I looked, and I felt the isolation acutely. This privileged little suburban undergrad, playing, it must seem to them, at sailing, while they, clad mostly in coveralls and grease, worked hard at keeping this vessel afloat and in forward motion.

I kept quiet and listened carefully and did what I was told, Ingo whispering translations in my ear.

Once the all-clear sound—we'd mustered satisfactorily—and the drill concluded, those of us who had nowhere to report milled about. "Come, sit," said Alois, motioning to the deck chairs. He had some nautical rope with him, and knotted it into intricate shapes. I'd never seen anything like it before; the complex patterns he created with a soothing motion reminded me of my grandmother's knitting. It filled me with wistfulness for home. I would have loved to ask him to make me a keychain or something, but I didn't want to open the door and let him think he could slip a toe in. I almost wished I were attracted to him. It would be exciting and romantic to have a beau on board, but it would also be stupid and complicated. Besides, the whole garden-gnome look didn't do it for me. The truth was that Roland—dark and brooding—was more my type, and he was too aloof and, probably wisely, too uninterested.

A colossal cloud way in the distance flashed lightning periodically as if it were trying to plug into some unseen outlet out there all alone in the slate ocean. Chris and Karl joined us, and it felt like we were sitting on someone's back porch. For just a moment, I forgot where I was; I felt like I was among friends.

I got up to go inside once the sun had dipped all the way down. Alois said, "Let's go up to the bridge. You can see the stars." I suspected his offer was nothing more than another attempt to strengthen our tenuous thread of connection, but I'd longed to stargaze ever since Herr Rose had invited me.

"If Herr Kapitän or Herr Betz are on duty, they won't want me there," I said. Betz always baited me when I cleaned in the morning, and I studiously avoided him.

"I don't know who has watch. We will go up and find out," he said.

Fortunately, it was Herr Rose, who seemed pleased that I'd taken him up on his offer. The three of us walked over to lean on the railing before the wide windows. The dark was so dense that it engulfed the slow bounce of the containers below. The sky was clear now that we'd left the lonely thundercloud behind.

"Let's go outside. We will see more clearly," said Herr Rose, and he opened the door to the bridge's side deck that protruded over the ship's edge, way above the deck below where we had just been sitting, and just above the deck near my cabin. The sky was a navy-blue, pastina-dotted soup. A celestial riot compared to anything I'd seen, even up at the lake house on a moonless night. There was no sound. No scent. Just sky.

"Now that we have passed the equator, you can see the Southern Cross," he said, pointing up. "It is not possible to see it in the Northern Hemisphere."

Alois added, "And there you see Scorpio" —he outlined it with his finger— "and there alpha and beta, the two stars closest to the earth after the sun."

We went back inside, and Herr Rose explained how, even in the absence of satellite guidance, they could chart our course provided they had a clear view of the sky.

I felt like Galileo. As long as I could see the sky, I wouldn't get lost.

At the same time, each milestone we passed en route to Australia took me further from home. We'd left the East Coast, but at least I could understand the Spanish in the canal. Once through the canal, we dove into an ocean I'd never experienced, and my connections to home felt even more tenuous. I forgot how their voices sounded and could not easily conjure faces without looking at the few photos I'd brought. Crossing into a different hemisphere turned me upside down. I was gone. It filled me with as much longing as wonder. I felt the separation more acutely, and wondered how I'd stay attached now that the equator had even more definitively severed the cord. I was suddenly untethered and ungrounded, and it had nothing to do with the ship's motion.

"Thank you," I said. "I'd better head back down."

Alois objected. "No, no, the night it's young, we can go back below."

But I'd suddenly had enough and just wanted to sit in my cushioned, carpeted solitary container.

Which surprised me because I spent most of my waking hours there. The work, no matter how Herr Most strove to find more for me to do, barely filled five hours each day. Since I started most mornings at 0600, I finished before lunch. Even meals took up only so much time. No matter how much Ingo or Karl engaged me—and I appreciated it when they did—I never lingered for long because I felt too self-conscious, given the palpable air of antipathy and indifference that wafted from the others along with the scent of spaetzle. How resentful they must have felt about me spending my summer vacation at their place of work as a favor to some American shipping agent in New York!

I would lock the door behind me and let my key and my shoulders drop. I tried to keep to some kind of structured routine to minimize the pining and make the hours pass. I missed my family and friends. I missed all things familiar. I missed an environment that wasn't always mildly menacing.

I'd put on an eight-track and let the music reconnect me to home, as I sat at my table, treating it like an office desk, and wrote in my journal and composed letters home. I'd sent a few from the canal, and would have a slew to post in Sydney. I started writing poetry at that table, inspired by the natural wonders that surrounded me every day beyond the ship's railing: the vast sea, the fickle sky, and the sun and the moon that rose and fell with symmetrical precision in infinitely varied iterations.

My grandfather, or Papoo in Greek, wrote poetry, too. When I got home, my father's eyes welled up to see that I'd followed in his father's footsteps. He bound both of our oeuvres in matching hardcover volumes that I still have on my bookshelf. I doubt the verse was good, but it did help me spend hours in creative reverie.

I would eat my schmoke-time treats at the table, too, taking care to set it up like a proper meal so it would take more time. I cut the pastries with a fork and knife, and chewed slowly like Mom told me to. I tried to stretch out every activity in my cabin as hard as I tried to complete the manual labor outside it quickly.

I organized and reorganized the few things I had on board. I cleaned the cabin often. How amused my mom and my college roommate would be. Mom had cajoled and threatened me to get me to clean my room at home. At school, Randy and I would let the dust bunnies accumulate under our beds until we feared we'd have to start feeding them. And here I happily took the supplies Herr Most offered me and vacuumed, dusted, and cleaned the bathroom as if I'd expected my next-door neighbor, the Herr Kapitän, to stop by, donning white gloves to check for grime.

And I read, either stretched out on the golden sofa, or curled up in the well just above it, where I would gaze out of the porthole to let the omnipresent but ever-changing sea and sky hypnotize me. By then I'd made it through Hailey's *Airport* (my dad gave me that one) and was well into Updike's *Rabbit Run*. I longed for someone with whom I could discuss that dark novel. I sensed that its themes lay just beyond the reach of my age and experience. On the shelf, *Slaughterhouse Five*, *Women in Love*, and *Watership Down* awaited. I knew that once I got to Australia I'd have to buy more. When I packed, I'd had no idea just how many hours I'd need to fill.

I made feeble attempts to exercise, but resisted the temptation to do laps in the small pool because I did not want the men to catch me clad in a bathing suit. So I feebly sat up and pushed up to try to keep down the amount of fat I may as well have injected directly into my cells in the form of pink strawberry pastry cream filling.

I tried, especially on nights without Kino, Kanteen, or some other activity, to just go to sleep early. The sheets were soft and the dreams of home sweet. But even with a 0530 wakeup time, sleep eluded me at 2030, no matter how bored I felt. The vast oceans of down time left me adrift between exhaustion and sleep.

I tried not to think too much about what everyone at home was doing; all those things that, as a smug college know-it-all once seemed so plebian. I longed to sit with Lauri and Sharon and sip too-sweet iced tea all afternoon; with Sharon gone that would never happen again. Or ride my bike into town with them to sit at the Woolworth's counter for a lunch of Coke and fries, the former served in a white paper cone cup in a red plastic hourglass-shaped base, the latter in a white wax-coated paper box with a red-lattice design to make it look like a basket, and, I suppose, to match the soda cup. Or if we could convince one of our parents to let us use a car, we'd drive out to the Howard Johnson's near the Garden State Parkway's Exit 135, and annoy the waitresses by ordering a carafe of coffee and nursing it for hours while we talked and talked.

I even missed the things I thought I hated. Like my heretofore annoying younger sister. Herr Stuhlemmer let me call her on her birthday from his office near the bridge, but the crackly connection and exorbitant expense prevented me from talking to her for long.

Or like my father's bad chicken jokes and tendency to control . . . well, everything. Shortly after I arrived at Middlebury my freshman year, I found a letter in my mailbox. He must have mailed it the day he got back from dropping me off, if not before we even left. He, with nothing but the best of intentions, had pored over the course catalogue and constructed a four-year plan of classes that he "thought I'd like to take." Did he mean him or me, I wondered? But that letter screwed me from day one. If I followed his suggestions I'd major in something "practical" that did not interest me. If not, I'd implicitly disappoint him. I convinced myself that majoring in economics was really my choice, but he had made it for me the day he sealed, stamped, and sent that letter. I crammed in all the English and Spanish literature classes I could, tucking them in any crevices left between the economics requirements. I dared not defy Daddy. Yet even so, I could not wait to see his unmistakable ornate handwriting in letters I hoped would meet me in Sydney.

Or like my mother's loud, shrill voice and her soft, compassionate heart. No one takes care of you like your mother. Especially no one on board a German container ship full of men in the mid-Pacific.

But also, I was determined—hell-bent, really—to see this adventure through. In some ways this was another "class" my father had urged me to take, but I wanted the lessons: to learn to be alone with my time, my thoughts, and myself, and to learn how to cope with isolation and scorn. Because even as I was once again following my father's direction, I was striking out on my own, like Captain Cook, to explore the land Down Under. Or so I told myself. Over and over. The seeds that Dad had planted, and I watered with compliance, would produce a very different blossom than he'd imagined.

———————

The phone rang one mid-Pacific night at around 2100 hours. Tim invited me down to his cabin.

"I want to talk to you," he said. "Nothing funny, I promise. I just need to talk to you."

I hesitated for a long time. "I am so tired," I said, even though I couldn't sleep. "Does it have to be right now?"

"I will not keep you long, I promise."

The lowest ranking crew cabins sat well below deck. The air was so heavy and humid at this depth that we may as well have actually been under water. I could taste the salt water as I breathed it in. Even at night, the moon and outdoor fixtures lit the cabins above deck a bit. Down here, there was only the gloomy glow from metal-caged bulbs.

I tapped on Tim's door, which felt like it was coated in something between mist and slime. It was lit even more dimly inside, so that everything blended into a denim-blue murk. Must and dust mingled with the scent of stale beer and cornflower-blue coveralls that needed laundering.

With one foot in the hallway and one in the cabin, I saw Tim's outstretched hand. Ostensibly a gesture of welcome, it felt more like Charon welcoming me onto the ferry across the River Styx.

I was diving into a dangerous swimming hole, deep in the backwoods, forbidden by all parents to their children. "Come on in," he said as he took my hand in his damp one. *The water's fine*, I imagined I'd heard.

"Thanks for coming down," he said.

"Sure," I mumbled. *Sure?* I asked myself.

"Here. Sit," Again, the imperative. I could barely see the place to which he pointed. Was it his bed? I sank into something dark.

"You know there is tension," he began, "With the guys, I mean."

I nodded. I knew. Tim and the puppy-like Chris had been at odds for a while, and Roland had been pissier. I just assumed that Chris's cheery, chirpy demeanor annoyed Tim and that, well, everything annoyed Roland. I wondered why Tim felt the urgent need to share this with me. What it had to do with me.

"It's over you," he said, an octave lower, for dramatic effect.

"Me?" I asked meekly. Shit. I didn't want to hear this. I wanted to be back up where I could breathe. Alone.

"Yeah, you. We want to know. . . ." he trailed off, clearing his throat, his blond, stringy hair obscuring his face as he struggled to come up with the right words, perhaps, even though English was his first language, too.

"We want to know which one you choose."

"Which what?" I ask dumbly.

"Which of us. Which one you want to sleep with. It would be easier if you just tell us so we don't need to fight about it and get annoyed with each other. Me, Alois, Chris . . . We gotta work together, you know? Kapitän Beucking made it very clear to us before you came on board that we were not to bother you. That we were to stay away from you. Nothing against your will. If you tell us who you choose, it will be easier."

I was dumb. Dumbstruck. Dumb not to have seen this coming. Dumb because there was not a word my mouth could formulate. All I could think was that Tim was married and had a kid, and that I did not want to deal with this shit at all. And how could I get the hell out of there quickly?

He'd cornered and trapped me like a big cat taunting a small mouse. I knew my answer could determine how the rest of this long, strange trip would play out.

"No one, none of you," I stammered. "I have a boyfriend," I added, as if that would close the case. The naiveté of this defense, in light of the fact that this married father was the one posing the question, startled me even as I stammered it out.

Tim looked down at his hands and shook his head almost imperceptibly. This was not the answer this chosen emissary had hoped for.

"Tim, I just want to be friends with everyone." I now oozed naiveté like a squeezed tube of toothpaste. "I'm sorry," I added, completely uncertain what I was apologizing for as I gingerly backed out of his cabin and retreated to my own to lick my wounds and ponder my fate.

Despite our proximity to the equator, a chill fell over the ship. Tim must have immediately broadcast the results of our chat. Those in the running and those who had placed imaginary—or real, for all I knew—bets on the outcome froze me out. Except Alois. I guessed hope sprang eternal in him, and he saw no downside in continuing the chase. Also, as an officer, he had a different mindset and position on board and in the company than the others. Steadfast allies, like Ingo, Karl, and Herr Most, stuck by me. I had no way of knowing if they'd heard what happened, but they remained as amiable as ever. I was even more grateful for their small kindnesses than before. But the excommunication took my isolation to a new level. What before ranged from indifference to distain ratcheted up to thinly veiled hostility. One man, in particular, a greasy-haired machinist, would glare at me so intently that I felt sure he was imaging what he'd do to me if he could.

That slut! I imagined them saying, as they glared at me from under furrowed brows. *Aren't German men good enough for her? What a tease!*

I had studiously avoided flirting with anyone, dressed as conservatively as I could in the heat, wore no makeup, kept my hair tied back, and kept largely to myself. I resented being so confined, and resented being put in this position despite my virtually pristine behavior. I had, in fact, "gotten myself to a nunnery," feeling even more sequestered than before.

I was the odd girl out in the school cafeteria, the target of mean-girl gossip. But these mean girls were big Aryan men.

I kept a low profile and felt even lonelier. I had no choice but to work at Kanteen, and I'd slip in and out of Kino night just because I was desperate for something to do. They showed *The Fortune* with Jack Nicholson, Warren Beatty, and Stockard Channing, in English with German subtitles. The English sounded so soft and mellifluous compared to the hard-consonanted German that surrounded me. But I no longer attended any other gatherings unless they were ship-sponsored, and no more of those were scheduled until our return trip. Even the waves pawed and groped more menacingly as we dove deeper into the South Pacific. The rhapsody in blue was turning into a cobalt nightmare.

Herr Most must have caught wind of what happened because he treated me a little less curtly for a few days, and stepped up the "don't let the boys take advantage of you" rhetoric. Yet he never directly asked how I was, what I was thinking, or about my experience on board. He was my boss, not my therapist.

One morning he thrust a few oranges at me while scrutinizing the cuticles on his other hand.

"Here. You must have vitamin C. You make finish early today. We pass Tahiti. It is nothing, really, just a big island, but if you want to look go upstairs, *ya*?"

I thought of stories of sailors in the olden days coming down with scurvy due to a lack of vitamin C; at least I'd avoid that fate. He knew very well I couldn't wait to see Tahiti. I'd been asking about it since we left the canal.

I looked up and took the citrus. This was kind and unusual treatment. I just nodded. "*Danke*." I didn't want to say too much more because I wasn't sure the words would make it up past the lump in my throat.

I perched on the smaller, more-secluded deck up by my cabin, alone and slathered with sunscreen this time. I took in the soaring, green-velvet spires, cooled by a tropical breeze. They were like jagged, spiky canines of some monstrous sea creature that hadn't brushed his teeth, ever. I could just barely make out beige strips of pearly beaches below those majestic peaks, and I imagined happy, bikini-clad women frolicking with their new husbands in the clear cyan South Pacific lagoon. I longed to be on that shore in the shadow of the lush foliage. But soon the emerald crests began to shrink as we sailed past, fading with the fantasy.

Even with my new dawn duties on the bridge, there was just not enough work to fill eight hours. Herr Most wracked his brain to concoct chores, but he would not entrust those jobs with direct officer contact to me, nor would he assign me the arduous and dirty crew jobs. On most days, he would say "You make finish, *ya*?" long before lunchtime.

The weather was fine, so this was fine with me. The sun blazed unimpeded except for those few days when we'd sail into and quickly out of a refreshing rain shower. I'd run down to eat whatever lunch was on that day of the weekly meal rotation—in ever more abbreviated stays in the mess—and then head up to what I viewed as my own private deck. Just down the hall from my cabin, this small "wing" of a veranda protruded slightly over the other decks below. All outside flooring featured a mix of fern-green paint and something grainy that provided traction when navigating the often-slippery surfaces.

My small sanctuary was empty except for two very large, weathered teak armchairs. A larger deck, adjacent to the bridge where we'd stood to stargaze, jutted out just above me. The captain or another officer would often stand there to observe as a tug or pilot boat led us into or out of port. Occasionally now, the shift officer would wander out, I imagined, just to get some fresh air or look at something on the horizon. But out that far at sea, the horizon remained infinitely empty. I could assume almost complete privacy up there.

The sun pinged off the relatively calm water like sonar. The maximum forward velocity of twenty-two knots always manufactured a breeze at this level, as if an immense, invisible fan sat poised on the bow, blowing back at us. But this close to the equator, the wet air weighed me down nonetheless.

In short-shorts and a halter top, and slathered in Bain de Soleil, I brought D.H. Lawrence up to share the afternoon on deck with me. I wrestled with the heavy teak chair to create a perfect vector between the sun's rays and the breeze. When I thought I had it exactly right, I dragged it just an inch further to get close enough to the white metal railing to rest my feet on the second rung.

I settled in and looked out to the horizon. The taste of salt on my tongue made me think that one of those hoppy amber Holsten beers would make this moment sweeter. No one could see me here. I let my guard and my shoulders down for a moment, and opened *Women in Love*.

Before I could dive into D.H., the door behind me swung open, crashing steel on steel, as it ricocheted off the outside wall. A heavy metal echo rang in my ears.

I heard him before I could see him, because although I turned quickly, he had started shouting even before he'd stepped out. Werner or Wolf or something—that lurking, low-level machinist. I didn't know him. I'd only seen him creep in and out of the mess hall. Of everyone on board, he scared me the most. Nearly six feet tall, with a well-developed beer gut, he mostly skulked and leered. I'd never seen him interact with anyone, nor speak above a mumble. He wore tattered, gray coveralls that I'd likely mended, and clunky, rubber shower shoes with white socks. He now raised and waved his bratwurst arms overhead, shaking his fists at me like an angry bear. He loped toward me, shouting something in German that I could not possibly understand.

I dropped the book, jumped to my feet, and backed up to the railing in the only defensive position I could muster: a vague cowering. My right forearm rose instinctively and involuntarily overhead, and I held on to the railing with my left.

He charged toward me and continued the incomprehensible tirade. I knew with absolute certainty that he would pick me up in his large claws and throw me overboard. Would I make it to Australia or back to Middlebury? Would I ever see my family again? My heart rate was so high that blood would surely spurt from my ears at any moment. Herr Rose's words echoed loudly: "Don't go overboard. We would not turn the ship around to fetch you. You would be dead by the time we could reach you."

But he paused for a moment, like a beast distracted from his prey by a butterfly, took one step back with his right foot, grabbed the teak chair in both hands, and held it high overhead, like an orangutan waggling a limb overhead to assert his dominance. I'd moved the chair around enough that I knew it was heavy. But he shook it and shook it like a baby rattle, as he continued to shout with such rage that hot spit hit my face like spewing sparkler embers.

So he's going to throw the chair over, and then me? Instead of me? We seemed momentarily to be mulling the ramifications of this choice together. And then, just as suddenly as he had burst, a bucking bronco, out of the door, he stopped. His face looked like it would boil over and his head would melt as he slammed the chair back down. Not once, or twice, but three times. The deck below me reverberated with tsunami waves after each contact. He slammed it down with a thud, and punctuated his rant with one final point. I considered myself very lucky not to be able to understand the slew of invective that he hurled at me. He turned and stomped off back to his cave.

I stared at the door that closed behind him for a long time, completely unconvinced that he'd not just gone inside to grab an axe or some other sharp object to finish me off, even though I could not imagine what I'd done to deserve this.

When I finally started breathing again, and could feel my body at all, I realized my knees had buckled and I had crumpled to the floor with my feet tucked under me. Every molecule of every bone, organ, and inch of skin vibrated so violently that my teeth chattered and I thought I'd throw up. I let my head drop near my knees because I felt dizzy with hyperventilation. And only then started to weep. The sobs wracked my body in time with the concert of everything else that was shaking.

I want to go home. I want to go home now. Right now, was all I could think.

I lay there, contained in myself, for I have no idea how long. Until the quaking quieted. I reached for the book and stood tentatively on uncertain fawn legs to make my way inside.

I cracked the door open ever so slowly, half expecting to find him waiting inside to finish me off, and crept to my cabin. Behind the bolted door, I turned on the shower, sunk to the floor of the stall, and curled into a ball to let the water wash away the snot and the salt in my hair. I'd settled down into a post-ugly cry-whimper, glad the water would drown out the sounds. I tightened into myself and just let the cascade quiet my mind and wash away the image of him wielding the chair overhead and the sound waves of his harsh, incomprehensible tirade.

I had to collect myself because if I didn't go downstairs to collect my snack, Herr Most would come looking for me. My eyes were as puffy as the pastry he had on a plate awaiting me. I kept them down, as if avoiding eye contact would somehow prevent him seeing the distress. I needn't have bothered. His own gaze down, fingernail-focused, he said, "So. What happened, Fraulein Meyer?"

I started to cry, as furious at myself for not being able to hold it in as I was at the raging machinist.

"I don't know! I have no idea!" I said, telling my truth. "He came up screaming at me. I thought he was going to throw me overboard."

"*Ya, ya*, Herr Betz heard everything from the bridge deck. All the men on the main deck, too." Oh great, the adversarial officer and the public at large heard it all.

"I didn't understand him," I said. It began to dawn on me that somehow *I* was in trouble.

"Ya, ya Fraulein Meyer. Herr Wimmel, he works at night, so he sleeps at the day. His cabin it is just below the deck where you sun—."

I interrupted. "I was reading!" but he stopped looking at his nails long enough to put his hand up. I froze.

"So when you move your chair you wake him up."

Oh my god, I thought. I understood. While maneuvering the massive chair around to get it positioned just right, I was jack-hammering his ceiling.

"You should apologize to him."

Wait, what? *I* should apologize to *him*? I didn't realize what I was doing and certainly didn't mean any harm. He, on the other hand. . . .

"*Ya, ya* Fraulein. Herr Kapitän is furious with him. I am certain he will discipline him, which will not make him happier. He was wrong, but it will help if you can say you are sorry."

I understood that he was defending me and not accusing me. He was trying to prevent further animosity. I did not want to cause trouble for my boss.

"I understand. I will say something to him at dinner."

He nodded, and refocused on his fingers. The discussion was over, with nary a nod to how it might have impacted me. I took my treat and retreated to my cabin, humiliated, frustrated, and not just a little scared.

At dinner that night I walked up to his table on shaky legs. My head was down, as was his. He looked more like a chastised little boy bully, despite the five o'clock shadow.

"I am sorry that I woke you up. I did not know your cabin was below that deck. I did not know you were sleeping," I said.

He nodded. I do not think he spoke one word of English, but he could certainly tell from my demeanor that I came in defeat and not offense. All eyes were on us, and I felt I might melt, Wicked Witch-like, into a puddle right where I stood. That was it. He nodded. He didn't say a word. Didn't apologize to me. Perhaps he didn't have the words in my language, but just as I'd conveyed my meaning, so too could he have. But he didn't. I slid back into my seat. The show's over, everyone, just eat.

Land Ho

June 29, 1979

33.8688 S, 151.2093 E

Rumors flew around the ship like the odd white birds that circled us. We wouldn't get into Sydney for a week, they said. Dockworkers in Australia had strikes and slowdowns often. It could take days to unload and load once in. The talk swirled with the birds. The closer we got, the louder it got. With every squawk I calculated and recalculated our arrival date back into New York in my head.

"That would put us back on August 27," I said to Ingo. "I'll never make it back for the start of school." He smiled and wiped his hands on the apron that struggled to circle his belly.

"Ya, ya. They can start without you I'm sure!" He nudged my shoulder. My distress amused him. I should have fretted to myself.

On June 29 we finally got close enough to see the big island continent. "Land ho! Land ho!" I wanted to shout from the nonexistent crow's nest, like a crazed pirate too long at sea.

We anchored and waited. A wash of terra-cotta-tiled roofs dotted the hills by day; at night the lights ringed the curve of the coastline like strands of diamond clusters. Like Pip longing for his beloved Estella, I saw Sydney as "tangible yet unattainable." It was almost too much to bear after seventeen days of ocean, to see it, to imagine my toes touching an unmoving surface, my ears hearing proper English, and being understood rather than ribbed and ridiculed, but still to be stuck on the ship.

That night at dinner, I asked Karl if he knew the name of the birds that dove like bombardiers around the ship all day.

"They're not seagulls," I said.

"No, they are . . . let me think of the name in English." He tugged at his tight blond curls. "Albatross," he said.

I laughed out loud.

"What? Why is that funny?" he asked.

"Oh, nothing. There's this famous poem, *The Rime of the Ancient Mariner*," I said, "by Samuel Coleridge. A sailor kills an albatross, and they force him to wear the carcass around his neck as punishment. We have a saying in English about a burden being like an albatross around your neck."

He mulled this over for a moment. I could tell he was translating and trying to make sense of it. He kept twisting his hair as if it helped him think. Finally he laughed. "I see. So perhaps we are your albatross?" He laughed again. I didn't, even though I knew he meant no harm.

"Or perhaps I am yours." I headed up to await our grand entrance into Sydney Harbor in my cabin where I could curl up in my window well and watch alone.

Just as I slipped the key into my lock, the captain opened his, startling me. In all the time on board, we'd never bumped into each other at our doors, even though they abutted. When we did meet he only barely nodded in acknowledgement. He'd scarcely said a word to me, and I suspected he did not welcome my presence. Women on commercial ships caused problems, especially young unmarried ones. He was still in uniform, but he'd undone his white shirt's top button and donned brown leather slippers. I could see just past him into his cabin, and make out a sitting room separate from the bedroom behind it. A full suite. He looked down and a little flustered. I realized he was not so much aloof and arrogant as painfully shy. A man of middle age, no wedding ring adorned his finger; he seemed a loner. He held something that I couldn't quite make out in the hand not on the door handle. He looked up, eyes only, barely lifting his head.

"*Abend*, Fraulein. I hope everything goes well for you on board," he said, so quietly that I had to strain to hear, especially through the accent.

"*Ya, ya, danke*," I answered.

He held out a book, offering it to me. "This guide to Australia, it contains all the ports we will visit. Perhaps you would like to have it for the visit. You plan to go ashore?"

"Yes, I look forward to it. That would be very helpful. Thank you so much, I really appreciate it." *Stop gushing, Diane.* "I will return it once we leave for New Zealand."

"*Nein, nein*, please keep it. I have read it. I have been here many times before." And he looked back down. "Well, *gute Nacht*, Fraulein. Enjoy your visit in Australia."

"*Gute Nacht*, Herr Kapitän, and thank you." He nodded again and turned to go into his dimly lit quarters. I cherished that book because of the kindness it represented.

I slept fitfully in the ship's stillness. The lullaby of motion silenced the ship, and I filled the void with anticipation and anxiety and excitement.

In the morning, Herr Most seemed unusually cheery. "We don't go in until July ninth," he said, looking right at me, not at his hands, almost smiling.

Oh no. No. No way. I tried frantically to do the math in my head, which I shook back and forth. "How can that be? Why do we have to wait so long?" He just shrugged his shoulders and concentrated again on his fingers. I stormed out and went to breakfast, too upset by the setback to even ask if he had anything for me to do later. This was devastating.

I plunked down at my place with my head hung low. I

nodded at Ingo, but just barely. He wiped his big hands on his aproned waist, and then came over and poked me gently in the shoulder with one of his pudgy fingers.

"*Was ist los*?" he asked. "What is wrong?"

"*Nicht*. Nothing. I just . . . Herr Most told me about the delay until July ninth. That we will just sit anchored until then. I'll never get home." God, I needed a fairy godmother.

Sometimes, though, they don't come clad in sky-blue crinoline. Mine held his shaking belly as he laughed loud and hard, more Wizard than Glinda.

"What?" I asked. "Don't make fun of me now, Ingo, please."

"*Nein*, *nein*, I don't make fun of you. Herr Most makes fun of you. He plays a joke! We go in today, around ten hundred hours." His laugh was a magic wand that made Herr Most's evil joke disappear.

That asshole! I couldn't decide if the potential delay or the arrow that Herr Most had sharpened and shot directly into my tender Achilles' tendon hurt more. That asshole!

We started in at 2220, sailing under the famous Sydney Harbour Bridge, known as the Coat Hanger, its suspension cables concave compared to the Verrazano-Narrow's convex ones. I felt no foreboding under this elegant arch. Its twinkling corona of light crowned us with the promise of land and egress. We passed the iconic Sydney Opera House, its huge crustacean claws rising out of the black harbor, reaching for the sky. The tiny mighty tugs did their job and had us tethered alongside the dock under the cranes in—relative to the crossing we'd just made—no time.

I was too excited about the arrival to sleep, so I wandered around to see what would happen. Two groups came on board in every port, regardless of what time we docked, each to service the ship in their own way. The first represented the port: two uniformed customs agents, not much older than I, sauntered through the hallways. I guessed that they needed to collect some sort of documentation from the captain, but also to make a visual inspection of the premises. For me, they were relief in the form of English-speaking, amiable eye candy.

"Hello, luv." Luv? I was smitten, especially with the accent. They removed their inspectors' caps. After introductions they asked the anticipated question: "Whatever are you doing here?" I gave the standard response.

We chatted for a while, and they told me where they came from, what weather I could expect, and made a few suggestions for things to see in Sydney. I wished they could serve as tour guides. I couldn't wait to get out and explore.

"You'll want to go over to the Opera House—it's beautiful and something's always going on there in fine weather," said one.

"And go wander around The Rocks a bit," said the other. "The prisoners from England landed there. There's a great place called 'Pancakes on the Rocks,' too. Good food and quite reasonable. And don't forget to taste a Fosters!" They both laughed, but with no ominous overtones, nothing but warm Aussie welcome.

I could have listened to them say anything in their accent for the whole evening, but they had to get back to work.

"Ta, luv." Love. Just what I missed and needed the most.

The second, less official group arrived to service the men. A gaggle of prostitutes always materialized whenever we docked. Not knowing their profession the first time I saw them, I wrote in my journal that night that "the ugliest group of girls I've ever seen just came on to the ship." Scantily clad, garishly made up, and dramatically coifed, they entered as one colorful, cohesive, cackling coven.

They waved at me, the ugly duckling, and giggled. I flattened myself against the wall to let them pass. I just nodded in greeting, struck dumb by the sight of them, and wondered, naively, which one of the crew knew so many women so well in Sydney that they'd come to visit this late at night.

"They are whores." Said Herr Most the next morning. "They come on in every port. For party with the crew. They are trouble. I am married. I stay away. You stay away, too."

"You're married?" He'd never mentioned it before. We rarely discussed anything personal. He wanted to be neither my father nor my friend. Heaven knows I didn't need another of the former, but I was desperate by then for the latter.

"*Ya, ya.*" He waved at the air dismissively. No big deal. Nothing to see here. "She lives in New Zealand. In Wellington. We married two years ago. I will see her when we stop there." He gave no more detail and clearly did not want to talk about his personal life. I had so many questions. "You stay away from them. They are not good girls."

I felt like such an idiot. He clearly got that I hadn't caught on, but thankfully did not thrust his toe in the door I'd left ajar for ridicule. I sensed that they offended him too much to joke about. Maybe he felt badly about the decidedly unfunny misinformation he'd given me about our approach to Sydney. Maybe he was looking out for me. Was he this inscrutable with his Kiwi wife? It seemed like an odd arrangement to me. Neither of them was young, and they'd only been married a short time. I tried to imagine how he'd met and courted her, but didn't dare ask. He seemed devoted and loyal, and the "girls," as he called them, bothered him. He had no intention of glorifying them, joking about them, or letting me get anywhere near them.

I nodded and he said again, for emphasis, "You stay away from them."

———

Even on a vessel as large as the *Columbus Australia*, standing still and walking required constant calibration and recalibration of my center of gravity. Every time I picked up my right foot, the surface below it moved from where it had been when my foot left it, so I needed to find the floor and rebalance myself with each step. More pronounced swell required additional attention. It helped to touch anything built in—walls, tables, or anything else that would not sway as much as I did. This even happened sitting down or otherwise stationary. The ground or cushion or padding under me always moved, and my upper body stirred relentlessly, like a wooden spoon in risotto, to keep vertical. Even in bed, I felt like a gyroscope. This micro-adjustment became second nature, unconscious muscle memory. Which worked very well as long as I was on board.

As soon as the motion stopped—say, when I disembarked in Sydney, after nearly a month on board—my body insisted on trying to adjust to the terra that was quite firma. The resulting "sea legs" made me nauseous. I was seasick on dry land. I did not feel ill one day on board, but for the first day in Sydney, I was so queasy that I nearly threw up. My mind was thrilled to be off the ship, but my body rebelled.

I sprinted off in the morning as soon as Herr Most released me, guidebook and camera in hand, to explore Down Under. He warned me several times to follow the thick yellow painted path out from the ship to the road where I could find a bus toward town. Apparently it provided a safe route through the maze of machinery and vehicles in constant motion at the docks. According to him, the workers would strike if someone violated the sanctity of the walkway. I in no way wanted to take responsibility for shutting down an entire port, especially as this would delay my return to New York. I stuck religiously to that yellow path, like Dorothy and her colleagues. I shared her tenacity in wanting to reach home with minimal intervention from flying monkeys and other dangers. As promised, it led me past the maze of warehouses, cranes, and rail cars ready to cart containers away to their ultimate destinations. The dockworkers' stares could not bother me because I was too happy to be liberated.

I had three goals for the land portion of my trip: to get to know each of the six cities on our itinerary as well as I could in the short time I'd have in each place, to take in some arts and culture, and to find the perfect souvenir for the long list of recipients I'd complied on the crossing, as well as for myself. In some ways, I viewed the last as a synthesis of the first two. Shopping may seem a shallow, frivolous activity, but for me, supermarkets, department stores, and all the boutiques in between gave a good glimpse into a society's priorities and preferences. I enjoyed touching tchotchkes and watching consumers consumed with their daily routines. I relished the treasure hunt.

I dove headlong into each of these pursuits for the four days we had in Sydney: languishing on the quay in the shadows of the Opera House, eating savory pies and chips, and watching slices of people's days as they passed, scrutinizing them for those little idiosyncrasies that made them very different from but exactly like me. I spent hours shamelessly eavesdropping. I sipped Fosters. I found Pancakes on the Rocks, and although I felt very self-conscious about eating alone, I forced myself inside, knowing I'd never have this opportunity again. The cute customs officers had not steered me wrong.

"What are you doing all alone this far from home, luv?" asked the waitress. I explained, proud to have made it this far. "Good for you. You enjoy yourself Down Under!" she said.

I joined a walking tour of the Rocks where the guide pointed out the exact spot where the "POMs"—"prisoners of her majesty" or "prisoners of mother England" had disembarked. Their journey was even longer and far more miserable than mine. Having traversed the globe in relative comfort, I couldn't imagine their plight, and felt a bit better about my own ship.

A few older couples from England shepherded me around like mother hens with a chick. They gasped and gaped when they asked, and I explained, about my circumstances. The more people asked—and reacted with awe and amazement— about my journey, the more it occurred to me that what I was doing might actually be out of the ordinary and an important achievement. It made me feel proud and stronger than I ever had been. At the end of the tour, I sensed they were reluctant to let me get back on board alone, as if they knew something that I didn't about the return voyage. I sensed that they'd have liked to keep me with them rather than letting me go back to the big red beast with the bad boys. I assured them that I'd be fine.

The next day Karl and I took a hydrofoil over to Manly Beach. I was happy to have the company, and it was funny to cross the bay so quickly on such a small boat. After lumbering across the Pacific, this craft amazed us both as it lifted us just above the surface and hovered across the waves. We hung over the smaller hull and let the Sydney Harbor spray cover and cool our faces.

On the way back, we had a clear view of the *TS Columbus Australia* from a very different perspective. She looked quite majestic berthed there, regal scarlet crowned with her unique round white funnel. I spied my cabin porthole just under the wide, angled panes of the bridge windows. I couldn't believe I'd crossed the Pacific in what, from that hydrofoil's deck, looked like a very small space. I felt a tinge of pride for my ship sitting just under the Sydney Harbor Bridge. She had sheltered and protected me like a mother hen.

"Happy Birthday!" I said to Herr Most as I entered the galley the next day for schmoke time. The thin slip of ship news announced his big day. "We should put a candle in your strudel!"

"*Nein, nein. Nicht,*" he grumbled, not meeting my eyes.

"What will you do to celebrate?" I persisted, ignoring his indifference.

"I walk to Sydney after dinner to buy some German champagne. I drink it when I get back. You can walk with me." Not quite an invitation, but he looked up for just a moment from the nail he was so focused on when I walked in. "*Ya?*" he asked.

"*Ya, ya*," I said, thinking how well champagne would go with the kiwi fruit he had covertly slipped me since we'd made landfall, but more importantly, about the momentous nature of the gesture of inclusion. I didn't want to act too enthusiastic, or he'd change his mind.

We left the ship around 1900, when he'd finished with the officers. He still wore his steward's outfit. Even the Good Humor Man changes into casual clothes occasionally. Did he ever wear jeans? Sweats? I couldn't picture him in anything but black bell-bottoms, a crisp, white, short-sleeved shirt, and sensible black shoes. I'd given no thought about what to wear. He didn't care, and I wasn't as self-conscious with him as I was with the crew. But I did wonder how to act. We had never spent this much time alone together, and I didn't want to make him regret having asked me.

It was three miles to town, and I had taken buses from the gate of the port, but they ran infrequently in the evening. While Herr Most may have intended to treat himself to champagne, I guessed he'd never spring for a cab. Or maybe he just wanted the feel of land under his feet and the fresh air on his face after being cooped up on board for so long. From previous visits, he knew a liquor store that stocked the brand he wanted, so we set off, I to abet his solitary celebration.

I wondered what we must have looked like, this odd couple, race-walking together away from our big, crimson floating home. His gait mirrored everything he did: short, clipped, staccato steps. Nothing wasted. One very young, one much older. Not walking close enough to be lovers, or even father and daughter. We kept a respectable distance. What would people make of us? What did I make of us? As it turned out, it was very easy.

He said only a little more than he did in the course of any day, but it revealed volumes. How long he'd worked for Hamburg-Süd (thirty years). When he planned to retire (five years, to New Zealand, to live with his wife). I could not picture this man living off the ship.

I didn't say much, either. I just listened, afraid that anything I said might break the spell and shut him up. Mostly, we just walked.

June is winter in Sydney, but the air was temperate. Cool enough for a sweater, but not cold. We found the store—small and narrow, with bottles of all shapes, sizes, and colors crammed into the front window. Price tags in Australian dollars sat in front of each. It was warm inside, and my eyes wandered the shelves, looking for items familiar and foreign.

Herr Most asked for three splits of the champagne. One for tonight, one for the visit with his wife in Wellington, and one for . . . But before I could spend too much time wondering, he paid for the bottles with characteristic efficiency, cradled them in the paper bag with uncharacteristic tenderness, and said to me, "*Schnell*," as he turned to leave.

I rushed out after him like a puppy that has realized its master has moved on while he was still busy sniffing the ground.

We walked back in dusky silence, each of us adrift in our own thoughts, but together on our journey. When we got back to the ship he turned to me before we parted ways. I realized I didn't even know where his cabin was: down with the crew or up nearer the officers? He nodded, said, "*Gute Nacht*," and handed me the third split. But before I could even react, thank him, or wish him a happy birthday, he'd turned and walked briskly away.

Before we were to depart Sydney I rushed around stocking up on supplies: more books, an oversized journal for transcribing my poems, and some snacks and drink mixes. I bought a jar of Vegemite, even though I didn't much like it, because it so represented the country. I perched once more below the

Opera House's fanned shells to soak in the sights and sounds of my favorite spot.

As I was writing in my journal, a young man with a scruffy beard and a large backpack, engaged in the same sort of writing, turned to me. "American?" he asked. "I'm Peter. From Seattle."

"Hi, Peter from Seattle. I'm Diane from New York. I'm working on that ship," I said, pointing at her.

"Wow. You win! I've just come from Asia. I plan to spend some time traveling around Australia and New Zealand before I go back. I left a corporate job at IBM. I just needed to see some of the real world before buckling down in, you know, the other real world."

I nodded my head, even though I clearly did not know. "And what will you do when you return?" I asked.

"I don't know yet, but it won't be IBM. I'm sure of that. I suppose for you it's easier, knowing you'll go back to school." I did have a pretty clear path ahead of me for the next two years. He, not so much.

I nodded my head in agreement, but I knew that this trip would change everything. "Yeah, but I will feel very different in some ways, you know?" I replied.

It was his turn to nod. "For sure. I know exactly what you mean." Peter had given me permission to breathe again, to look at this trip through different eyes. I was self-conscious because I felt attracted to him. He must have been ten years older than I, but I just wanted to put my hand on his scruffy beard and pull him close in for a kiss. Just one kiss, so I could remember what it felt like. To inhale that male scent that wasn't tinged with grease, sauerkraut, and threat.

I hated to go; I hadn't realized how much I'd missed really talking to someone—someone who didn't seem always to be ridiculing me in some way—but we would depart that evening, and Herr Most had cautioned me to return well in advance of our estimated time of departure, even though we all knew we'd go later than scheduled. "You cannot always trust the timetable," he'd said. "We'd hate to leave you behind."

Peter and I exchanged addresses and wished each other safe travels. I wonder where he is now.

I made it back, careful to toe the yellow line, and entered the ship just as the half-dozen or so ladies of the night, headed home, definitely looking worse for the wear. "Hello again, luv!" they said in unison as they giggled.

"Hi," I replied, keeping my head down and my distance in accordance with Herr Most's edict.

But they surrounded me like an amoeba ingesting a smaller organism.

"You're adorable," said one.

"How old are you? You look so young," said another.

"Are you a Yankee?" asked a third, while reapplying bright red lipstick.

There was no way out but through, so I held my head up high and said, "Nineteen, and yes. I'm Diane. It's nice to meet you." *Ugh, nice to meet you. Really?* This was hardly a cotillion.

Divine intervention came in the form of a door slamming somewhere down the hall. The Hydra turned its heads toward the noise, then back to me, and tittered again.

"You're lovely," said yet one more. "Enjoy those boys! Ta, luv" And they chortled their way toward the shore.

I tried to reconcile the monstrous picture Herr Most had painted of these girls, heavily painted, but not much older than I, and really very sweet. We each, after all, worked on the ship. Just in different ways.

———————

Ingo had gone shopping, too, for fresh seafood, and prepared a feast that evening before we left Sydney. After a month on board surrounded by an ocean rife with life, I realized we hadn't eaten fish once. He bought huge pink shrimp and dark brown lobsters that didn't have claws like the ones at home.

And some fish that he cut into steaks. Even though we had a room full of Holstens below, he bought cases of Foster's oil-cans to give the feast a genuine local feel. He set up the same rectangular grills—they looked like oil barrels sliced the long way—on the main deck that he'd used for the "Western" BBQ. That night at sea, he and the other cooks had charred endless cuts of landlubbing mammals: bloody beef, lamb, and pork. Here, near land, he deftly seared gifts from the sea. He stood, apron-clad, behind the smoking coals, flipping seafood deftly. The jolly master of his domain attempted an Aussie accent as I approached with my plate.

"G'day, Sheila! Shrimp on the barbie and a Fosters to go with it?" he asked with a sly smile.

I couldn't help but laugh. The attempt at Australian layered over the German lilt resulted in a hilarious hybrid. Ingo could always make me laugh.

"Yes, Mate, *bitte, danke*," I mixed my linguistics as well, much to his amusement. The men were jolly and tame, sated in every way, which made for an easygoing last evening in Sydney and let me hope that the return trip might not be so bad.

We sailed out at around the same time we'd come in, and I watched the jeweled collar of the cove recede, wistful. I felt refueled, like the ship, in large part thanks to the warm welcome Sydneysiders had given me on behalf of this island continent. I watched until we passed back under the Coat Hanger Bridge's halo, and went back to my cabin to plan for Brisbane.

En route, on the Fourth of July, I alone celebrated America's birthday by sporting red, white, and blue: denim shorts, a white T-shirt, and red bandana holding my hair back. I felt a little silly, but also like I had a torch to bear. I couldn't let the day go by without notice, especially after all the ribbing I'd taken for my country and her myriad sins.

When I walked into the mess thus clad that morning, Ingo grimaced, grinned, and started whistling "Anchors Away" while saluting.

"Very funny, Ingo."

"Fraulein Meyer, you are so patriotic!" he said.

"I just stand up for what I believe." He kept whistling. Maybe I was celebrating my independence, too.

In Brisbane, Australia's "south" because it's further north and hence closer to the equator, the temperature rose with the sun. We had fewer containers to unload and load there, so I had less time to explore. This was the double-edged sword of the container ship. Cargo ships took longer in port—more time for sightseeing—but much lengthier round trips. We'd likely only have two or three days here, and all I wanted to do was get to the Lone Pine Koala Sanctuary that I'd read so much about for some marsupial cuddling. As soon as we docked, I marched dutifully down the omnipresent canary-colored path and out to the street to take several buses to await the ferry there.

At the park, I met the marsupials with another group of middle-aged English couples. They subjected me to the same inquisition and showered me with similar concerned parental protectiveness as my acquaintances from The Rocks. Our guide, dressed like Crocodile Dundee, complete with khakis and outback hat, led us around the forty-four acres of paradise. We saw Tasmanian devils and platypi, echidnas and emus, wombats and wallabies. But the kangaroos and koalas captured my heart.

Although I supposed I'd been drenched in nature for the whole voyage, this was different. I wandered through a wonderland of crocodile-green brush amid a court of kangaroos, roaming freely, many with joeys in their pouches. Any tension I'd been gathering and storing for six weeks just melted away as I knelt down with food for them in both fists, and stared right into the roos' big brown saucer eyes. They showed no fear, and just gamely met my gaze. They let me hold their front paws as they gently nibbled on kibble right from my palm.

"Lovely, aren't they, then?" asked one of the women, Edith, in our group.

I nodded, looked up at her, and realized there were tears in my eyes. They welled up and spilled out faster that I could wipe them away. I sat down on the ground next to my kangaroos.

"Oh, love, what is it? Are you okay?" She had her arm around me now. "Have you had a rough time on that ship? Have those men given you trouble? There, there." She pulled a tissue from her sleeve and handed it to me.

I wasn't sure I could even explain this flood of emotion. "I'm okay, thank you. Yes, I mean no, they're okay, and no one has given me a hard time, really. It's just been a long time since, you know, I've been home or had anybody be nice to me." I just wanted to hug her and the kangas and not let go. I don't think I realized how much I craved a little tenderness, and getting it from these endearing creatures and this lovely English woman had made my yearning surface.

The few marsupials that had gathered around me in search of snacks stayed and watched. Like Edith, perhaps they sensed my distress. We looked up at them with their little front paws held out and heads cocked to one side in question. We couldn't help but laugh. I couldn't keep crying with all these big, sweet eyes, including hers, staring at me with concern.

"There, now" she said. "Let's catch up with the others. They're going to see the koalas." I just wanted to sit there all day with my new friends, but when I stirred, their big muscular hind legs propelled them off, in cliché kangaroo style.

I hoped to have a hug from the koalas, but I had to settle for petting them. The strong, sharp nails that served them so well when scaling eucalyptus trees could go right through our skin, our guide cautioned. But we could get close enough to caress their rough, dense fur, and see the joeys in their pouches, too. They were either inured to gawking tourists or too stoned on eucalyptus, or both, to mind our proximity, and sat, Cheshire Cat-like, balanced where the thick branches met, while we fussed and fawned over them.

When it was time to go, I lamented leaving the sanctuary. I missed the unconditional love and cuddling that I'd found in abundance at Lone Pine.

Just as we got back to the bus terminal, mine pulled away. Distraught at the thought of arriving back to the ship late, I appealed for help to the driver sitting in the next waiting bus, its door wide open. "Hop on!" he said, and before I could explain that I just needed to know when the next one would leave for the port, he shut the door behind me and took off on an action-movie chase down Queen Street until we caught up to the other bus. As he drove, he chatted away so quickly in an accent so thick that I could barely understand him. I just held on and nodded my head. He would take no fare when he stopped to let me out. I thanked him and told him there wasn't a bus driver in the entire city of New York who would do what he had just done for me. "Ta, luv, you go on now and catch that bus!" he said, tipping his hat. I did, just barely, and when we reached the wharf, I dodged container-laden trains rolling in both directions to find the yellow path back to the ship. I arrived, breathless, just in time for steak and mushrooms.

"Busy day, Fraulein Meyer?" Ingo asked. I just stuck my tongue out at him. He chuckled and wiped his hands on his apron as he walked back into the galley.

The next morning, before I set out to explore town, I stopped by the galley. Herr Most rarely made me work in port unless he had something really important for me to do, but I always checked in out of respect. It was our little charade. He could have easily told me not to report the night before—to just take those days off—but I think he liked leaving me a little uncertain.

"What do you do today?" he asked.

"I'm just going to walk around Brisbane. Maybe go to a museum. Look for some souvenirs for my family."

"Hah. Souvenirs," he said, focusing on his pointer finger and thumb touching. "What do they need from here? It is all junk." He laughed again. "Before you go to waste your money, you stop at the Seafarers' Mission, *ya*? They have books and newspapers in English. You can get coffee."

I had noticed, but hadn't stopped in the one in Sydney. I wasn't sure what it was or if I qualified as a "seafarer." I felt odd about venturing in alone, but I realized that was a moot point given that I'd embarked on this whole journey alone.

"It's okay," he assured me. "You just tell them the name of your ship."

The single story building, its stucco like vanilla icing on a cupcake, sat just outside the chain link fence that separated the port area from the road. It was unassuming and easy to miss. The flying angel that protected all us sea folk hovered over the entry, and the plaque said, WELCOME. I entered tentatively, pulling the door open only a crack to peek in. It resembled the mission in *Guys and Dolls*, and I envisioned Marlon Brando greeting me in a double-breasted suit. It was also what I imagined a VFW club at home might look like. But instead of Jean Simmons, a small, prim, white-haired woman sat behind a large, old, wooden desk that made her look even more diminutive.

"G'day, luv!" she said, her bright blue eyes registering surprise. "We don't get many young ladies in here! What ship are you traveling on?"

"Good morning. The *TS Columbus Australia*," I said, trying to sound very official. "The big red one with the round white funnel," I added.

"Is that so? That's lovely!" she said, as she scribbled the name of the ship in her log. "Now that's an adventure, isn't it? How did you come to be so far from home at such a young age?"

"Yes, ma'am," I said, and explained again, for what seemed like the hundredth time, my circumstances. It no longer irritated me that people asked. I understood the curiosity. And each time I crafted the answer it felt more accurate, so that

I understood it better, too. What started as simply a way to fill the summer months had morphed into more. Into a test of my strength, a way to show my mettle. A chance to stand on my own two feet, however wobbly they might be at times, and to start taking steps toward independence on them. I was traveling to myself.

I spared her the esotericism, though, and stuck to the basics. She seemed delighted and came around to give me a hug and a tour of the facility, however unnecessary, since it consisted of mostly one large room that I could easily see from where we stood.

"And over here we have a little lounge area," she said, pointing to four forest-green, vinyl sofas forming a square around a laminate coffee table.

"Anyone currently working on a ship—or who ever has—can come in and relax. We have postal services, too, if you have anything to mail." Several older men reclined on the hardly comfortable looking couches, and read, chatted, or played cards.

"And there we have tea or coffee, and there is our library. Take a book or leave a book as you please."

"Thank you so much," I said. Herr Most had told me about the book exchange, so I'd come prepared with a Michener I'd just finished to leave, and perused the selection for new material to occupy me. She returned to her desk, and I noticed that some of the men were looking up at me and then at each other, askance. I suddenly felt self-conscious and a little out of place again, but they smiled and nodded, welcoming me to their fraternity. Anyone who walks through those doors, tested by the gods and protected by the saints of the sea, is a bona fide member for life. I belonged.

I selected a title or two, got some coffee, and sat down for a while. Although I was anxious to get into town, I didn't want to seem ungracious or unappreciative of the hospitality. I thanked the receptionist on the way out, and waved at the men, who sent me on my way with a chorus of "G'day, luv!"

I adored Brisbane, whose residents were as warm as the weather. I stopped into a museum briefly, but spent the better part of the day souvenir shopping. I found a plastic-sleeved copy of Roger Daltrey's *Ride a Rock Horse* and Be Bop Deluxe's *Moroccan Roll*. I was thrilled with these finds so far from home, and couldn't wait to show them to my friends at school, especially the ones at the radio station where I was a DJ. I hurried back to the ship, always just a little afraid that when I disembarked from the bus at the terminal, she'd be gone, subjected to some last-minute change in schedule. I sighed with relief to see her still tethered and waiting for me.

On the way to Melbourne, heading south, we sailed close enough to land to receive communication signals. I enjoyed "the best rock in Tasmania!" as we passed that Australian island state, while I mended linens in the lounge. I modeled my college radio shows on my idol, Alison Steele, the "Nightbird," from WNEW-FM, 102.7, in New York. I could hardly mimic her gravelly, sultry voice—I had neither smoked nor experienced enough—but I aspired to her cool delivery. I got a coveted ten-to-midnight shift, to which a lot of people listened while they studied. I played what was then current, but is now considered "classic" rock, in themed sets. I might play Roxy Music's "Angel Eyes" with Elvis Costello's "The Angels Want to Wear My Red Shoes" and The Rolling Stones' "Angie." Or Lou Reed's "Take a Walk on the Wild Side" with Springsteen's "Wild Billy's Circus Story" and "Wild Thing" by the Troggs. I was sure some listeners thought these combinations contrived, but it was sort of my trademark, and I amused myself concocting them. A small following tuned in regularly.

Now, sitting and sewing half way around the world from WRMC-FM, listening to a Tasmanian radio station playing a curious mix of American, British, and Australian tunes, I missed Middlebury acutely. I missed spinning LPs in that tiny studio with George and Lou. I missed eating bad food in Proctor Hall with my roommate Randy, and drinking three-dollar pitchers of watered-down beer at the Rosebud Café. I felt very far away, but knew that dwelling on it would only make me feel further away. I switched the radio off and tuned in Martina Navratilova battling Chris Evert on a tennis court on the television instead. It must have been Wimbledon. I just let the rhythmically bouncing ball pull me back into the lounge.

It was four days from Brisbane to Melbourne, where we would stay for only a short time since we had little to drop off or pick up, according to Herr Most. I was sorry because I especially liked that alluring city on the Port Phillip Bay, just north of Tasmania. It had a European feel, with its wide ornate bridges spanning the Yarra River. During my exploration, I spotted a theater whose marquis announced a run of Strindberg's *Miss Julie*. When the box office clerk said tickets were still available for that evening's performance, I bought one immediately, figuring I would just spring for a taxi back to the ship rather than taking the bus that late at night. My parents had given me a fixed amount of spending money for the trip, so I watched my budget carefully.

I'd studied Strindberg, Ibsen, and Chekov the previous semester with my favorite English Professor Bertolini. I couldn't believe the coincidence of it playing here now. I couldn't wait to see this Aussie production, and to tell Professor Bertolini about it when I got back.

It did not disappoint. I sat very close to the action in the small, intimate theater. Jean and Julie played out their struggle over sex and class. Her despair and demise devastated me. It made me think about my situation on the ship; how easy it would be to feel guilty because I came from a higher social stratum than many of the crew. And how easy it would be to acquiesce and sleep with one of them just to keep the peace and minimize the tension on the return trip. But, as the taxi drove on the left side of the night streets after the play, taking me back to my own drama, I dug in the heels of my flip-flops. No, I'd be damned if I was going to apologize for an accident of birth that I not only didn't flaunt, but made an effort to minimize. And I would be double-damned if they thought I'd sell myself like the gaggle of call girls just to appease them in exchange for some fragile détente. No. No razor blade for me. Fuck them all. Or not, in this case. I intended to keep my pride and integrity in tact.

The next evening before we left, I sat in the lounge and watched Aussie Rules football on TV. I wished I'd gotten to a live match; it baffled me. It looked like a rule-less melee, an odd lovechild of a *ménage-a-trois* between American football, rugby, and soccer. The announcers' accents mesmerized me as much as the players' mystifying muscled motion.

We left the Australian continent behind on July 12, and I watched the match until the signal faded, trying to hold onto the connection for as long as I could.

Friday the thirteenth dawned with foreboding. The sea, now tumultuous, tossed us as it had not before, and continued to do so for the four-day crossing to Port Chalmers on the south island of New Zealand. The dark, sharp-edged waves taunted the ship like young men with switchblades. Nothing stayed in place, including my body. Everything and everyone slipped, shook, and swayed constantly. I felt off, not so much from the motion as from the lack of sleep. Thanks to the bed's position facing the porthole rather than the bow, I never fell out, but the sea swelled strongly enough to pull me down to the foot of the bed and shuttle me back to the head all night long. If I wasn't moving, then the one item I'd neglected to secure or that had worked its way free would fall and startle me. All of us, even those most experienced, were a little bleary-eyed from the tumbling, and now in addition to steadying ourselves, we seemed always to need to reach for something that was about to fall, or to pick up something that had. I was juggling on a wobbly tightrope.

The worst was a day out of Melbourne when we came to a full stop. Without any forward motion, the ship rocked like a bathtub toy at the mercy of an energetic toddler. I held on to the galley's stainless steel counters to steady myself amid the frantic rock and roll, as I asked Herr Most, who seemed delighted to see my distress, "What's going on? Why did we stop?"

He stood steadily planted on the thick soles of those black shoes, rooted to the rubber floor, too tough to touch anything for balance. "*Nicht*. They need to replace something. In the boiler I think. We go soon, don't worry, you won't miss school! You look a little . . . lavender?"

"No, I'm fine," I lied. "And I'm not worried, just curious."

"Good. So from now we start to wind back the clock one hour every other day, so it will be darker in the morning. You start on the bridge at 0700 hours instead of 0600. Starting tomorrow," he said, almost reluctantly. I looked forward to the extra hour of sleep. He knew it, and thus hated it.

"*Ya, ya!*" I replied, unable to mask my smile.

One night Claudia fell ill, and Herr Most asked me to help serve dinner upstairs. I hoped she was okay but was thrilled to have something more to do. And relieved that he hadn't asked me to work downstairs; that would have been humiliating, and would really have given the boys a chance to have at me. I imagined Ingo and Bruno resentfully handling everything down there while I helped upstairs. I tidied myself up a bit more than usual and felt honored to serve the officers, most of whom had treated me cordially throughout the trip. They all dressed in their uniforms for dinner, and ignored me less than they had when I'd first stumbled into their space. They greeted me quietly or nodded, and thanked me genuinely. Their murmured conversation contrasted with the boisterous repartee and periodic taunting downstairs, and they took their time eating instead of shoveling the food in like coal into a boiler. The human contact and activity made me feel productive. It beat sitting in my cabin all night, wondering if I'd get home in time for school, and missing everything and everybody there. Herr Most seemed so pleased and relieved that I hadn't spilled anything on anyone or embarrassed him in some other way. It dawned on me that he might not have realized that I was actually fairly intelligent and responsible. He had no way of knowing that. And although I thought I'd proven it every day since I'd joined his team, this seemed to ice the cupcake of my competence, I suppose because pleasing the officers was of paramount importance to him. He gave me a whole bowl of yummy red cabbage salad and an entire carafe of green tea to take to my room, and added that I could take one any time. Now he tells me? As if I'd passed some test that I wished he'd given me much earlier. He also offered me a Holstens as he commented on my expanding waistline. Tempted though I was, I declined and retreated to my cabin, wishing I could serve meals every day.

Deciding that I needed a project to pass the oceanic hours during the trans-Pacific return trip, I'd begun to amass material for a collage of my experiences, including memorabilia like coasters, bus and attraction tickets, maps from each port, and images of the ship. I'd picked up several issues of *Cleo*, an Australian women's magazine, and planned to clip some pictures from them to include. I still had to gather raw material from the three ports in New Zealand, but I wanted to plan the layout. Karl had cut a three-by-three-foot piece of plywood for me to use as a base . I would look for glue and small scissors in Dunedin. I hoped that this, along with the poem transcription and the newly refreshed book supply, would keep me occupied on the crossing.

The next morning, after a particularly rough night, Betz had bridge duty. The ship swayed ferociously—the higher up in the superstructure you were, the more pronounced the movement. Every time I looked up and out at the vast expanse of now churlish, churning graphite-gray water, I could see the bow rise and slap back down like a whale's tail trying to quell the waves.

I watched the messy ocean, wondering when it would settle down, and halfheartedly polishing the burled control panel wood, when he said, "At this speed we go, you will never get home." Then he chuckled.

I stopped polishing, and looked down at the controls. Could I just push the lever to the proverbial "full steam ahead"? Would they make me walk the plank? Throw me in the clink? Thankfully, he did not see my despair because I had my back to him. I knew he wished he could.

"*Ya*, we will not arrive in New York until August twenty-seven by my calculation." He added.

Fuck your calculation, I thought, but I said, "Oh, that's okay. I will still have plenty of time to get back to school." Technically, I supposed I would, but it would leave me virtually no time to spend with my family or friends before I had to pack and return to Middlebury just after Labor Day. But what if we were delayed further? I did not want him to see my anguish, so I finished up quickly, left the bridge and fretted on my own. *Bastard*. Could I fly back home from Auckland? I had no clue how much that would cost, or if my parents would go for it. I hated to even let myself think how much the idea appealed to me. Just to cut the whole thing short and get off this ship and skip the return trip and get back to everything quotidian.

But I also had this strong sense that I needed to finish what I'd started. I did not want to cave just because the seas and the environment had become unfriendly. I kept thinking how lame I'd feel to tell the story of half a journey on a container ship. I dreaded disappointing my parents, not to mention burdening them with the extra expense. And how would I possibly get all my stuff on an airplane? I couldn't, so we'd have to meet the ship when it finally did dock in

New York to collect the rest. That would feel like such an immense, humiliating failure.

I held a pile of letters the shipping agent in Melbourne had delivered, mostly from my parents, each in handwriting so distinct that I felt I was holding their hands when I held the letters. Blue enveloped me. I knew that the high seas contributed to my low mood, and resolved to check into flight costs when I got to Dunedin and delay any decisions until then. I'd have to ask Herr Stuhlemmer to call them to see what they thought about the whole thing. I didn't want to think about making that call, or think about anything. So later I did what every good college student did when feeling overwhelmed. I got drunk.

We were close to the coast of New Zealand, and we were traveling slowly, which ratcheted up the rock and roll. After dinner Ingo and a few other guys took a cooler of beer and some rods and reels on deck to fish for sharks. I supposed if they caught any, Ingo would clean and cook them up for us. I'd never eaten shark, and had certainly never seen anyone fish for them off a container ship, so I sat with them to watch. I gratefully accepted every Holsten they offered me to assuage my angst. They got nothing, and I got a massive hangover.

———

Herr Most called in the morning and shouted in my ear that we'd land soon and he didn't need me. I was glad he wouldn't see me. I must have looked like shit. I felt like shit. I stumbled and fumbled to get myself up and out to the bus that went from Port Chalmers to the larger town of Dunedin, just inland. I almost missed it because I was in the small bank exchanging Australian for New Zealand dollars. I panted in gratitude as I tripped up the steps of the bus. The driver, smiling and patient said, "There's no hurry, luv. We will wait for you, but I'm glad you made it, because the next bus to Dunedin leaves in three hours." He shut the doors and asked where I was headed. I sat down right behind him and explained that I'd just come from the ship and wanted to explore town. "I'd love to find a coffee shop for some breakfast to start," I told him. He guessed that I was fourteen, and asked why I was so far from home. I wondered that myself.

The New Zealand accent sounded just different enough from the Australian to notice and was, I thought, a little softer around the edges, which is exactly what I'd end up thinking of the small island nation and its inhabitants, both human and wool-covered.

He said he'd happily point me in the right direction, and then told me there were more sheep than people in New Zealand. I finally settled back into the blue upholstered seat for the twenty-minute ride, and gazed out the big windows at some of the most picture-perfect scenery I'd ever seen. A lush gradation of jade green carpeted a slope that rose gently up from the sapphire-blue bay that we were driving away from. A handful of cotton puffs glued to the felt mountains by preschoolers came into focus as we passed countless lazily grazing, fluffy sheep. "It's so beautiful!" I said out loud, and the driver just nodded and smiled again, maybe seeing his countryside through my eyes.

That sweet man pulled the bus, about half-full of passengers, over at a coffee shop in the center of town and got out to escort me to the door. There was no bus stop there, and I doubted it was part of his scheduled route. He did it just for me. I was not sure whether it was his kindness, or the fact that not one person in the bus seemed upset, that astonished me more. I could just picture in my mind the revolt that would ensue if a bus in Manhattan veered off its trajectory to escort a nineteen-year-old New Zealander to a diner.

It may have been the only coffee shop in town, for all I knew, but it was perfect. Just what I needed to assuage my aching head and sit and think. I found a seat right in the big bay window that afforded me a panoramic view of Dunediners coming and going. I bathed in the flow of people, just letting the rhythm of their motion wash over me, in this land so distant and different from my own, going about their daily routines, just like we do: holding toddlers' hands as they cross the streets, greeting friends to exchange gossip, reading newspapers. The obvious differences—this small town nestled in a green velvet-lined teacup, the endearing accents—fascinated me, but the similarities amazed me even more. We all want the same things when it comes down to it.

And at that moment, we in the café all wanted coffee. I needed some food in my stomach, as well, to quell the quake.

"What can I bring you, luv?" asked the white-aproned waitress who appeared over my shoulder. She had short brown hair done up in a 1950s-looking pin curl style that reminded me of my grandmothers. She wore sensible, white, thick-soled shoes and had her pencil and pad at the ready. I liked her immediately.

"Coffee, white, please, and a cheese roll." I'd learned the jargon for ordering coffee with milk.

"Ta, luv," she said, and she brought it in no time. I eavesdropped shamelessly as I sipped from the mug, not to be invasive so much as to absorb the local flavor along with the caffeine. The coffee, maybe because I needed it so badly, tasted as good as any I'd had recently, and the roll was warm, spread with soft cheese; I presumed it had come from some of those sheep that I'd seen en route. I thanked them silently as I savored it.

Thus fortified, I set out to find my craft supplies and a travel agent. I'd seen the latter as I'd enjoyed my breakfast and the view of town. I settled into a blue vinyl, aluminum-framed chair across a wood-laminate desk from a middle aged woman who very much resembled the waitress, and asked about one-way fares from Auckland and Panama City to New York. Her raised eyebrow reminded me that this was probably not a routine request.

"Oh, luv, I don't hardly blame you for wanting to fly home," she said after I'd explained my situation. "Let's see what we can do for you."

I looked around at the posters that festooned the walls, enticing travelers to exotic destinations near and far—Rotorua, Fiji, Thailand—and thought how none of them appealed to me. All I wanted was to go home.

After a few minutes she gave me the bad news, which she graciously converted into US dollars for me: $845 from Auckland, and $270 from Panama City. That would be roughly $2945 and $950, respectively today. She looked apologetic. I looked dejected.

"Wow," I said. "Wow," I repeated, at a complete loss for anything more coherent to say.

"Yes, I'm afraid it's a lot. Would you like me to book a ticket for you?" She was asking to be polite. I knew she knew the answer.

"No, thank you so much. I will have to talk to my parents first."

She reached across and took my hand in hers. I fought back tears. "You take your time, luv. You will be fine."

I thanked her again and found craft supplies at the local Woolworths before I headed back to the bus that returned to Port Chalmers and the ship. I hoped she was right.

We stayed only briefly in Port Chalmers and the lovely south island, having little to unload. All the containers we loaded had telltale attached motors and fans indicating refrigeration. I suspected they were full of lamb, a major New Zealand export; Herr Most confirmed this for me in the morning.

———————

The rest of New Zealand went by quickly, which filled me with both joy and dismay. I wanted to begin the return trip as soon as possible, but I found the country enchanting, and would have enjoyed having more time to explore. I spent most of my shore days in the museums dedicated to the indigenous Maori culture. Tim even joined me one day, which surprised me on two counts: I did not think culture appealed to him, and it meant that perhaps his resentment toward me had dissipated with the miles since our chat. Shame on me for the former, and I kept my fingers crossed for the latter. As indignant as I still felt about his assumption and presumption, the voyage back would be immeasurably more tolerable if the leader signaled to the tribe that we'd achieved a delicate detente.

We met an old gentleman at the docent desk in the museum in Wellington. His blue irises shone from otherwise cloudy eyes, like sky peeking through an overcast sky. They belied the age that his deep wrinkles spoke of. It pleased him greatly to hear that we worked on the ship, and asked which one and where we were headed.

"I worked on the ships for many a year," he told us. "Truly, I sailed between here and San Francisco before they built either the Golden Gate or Oakland Bridges!" he looked past us now as he remembered. "Yes, it's true. I saw them both open." He winked and wished us safe travels.

I bought a small, dark carved wooden tiki with iridescent blue mother-of-pearl eyes for myself at one of the museums as a talisman for the trip back. I hoped it would smooth the seas and speed us along. By then, I'd put my faith in anything that offered even a glimmer of hope of removing obstacles in our path. Also, I found a very old walking stick for my dad. I'd missed Father's Day and his birthday since I was away; now I had one cane for each celebration. I hoped he'd love them. Thinking about him filled me with such mixed feelings. He loved me deeply and wanted the best for me, but to him, that mean better than *his* life, and as he pictured it *for* me. Even halfway around the world, I was desperately trying to please him. He returned or exchanged every single gift anyone ever gave him. Nothing was ever exactly what he wanted, and his own needs blinded him to the thoughts and gestures that went into the giving. Giving gifts to him became a joke in our family. We'd always attach a receipt (before the day of gift receipts) and include glib messages in our cards: "Happy Birthday, and hope you enjoy the gift you eventually get yourself!" But really, it wasn't funny. It was more important that he satisfy himself than that he make someone else feel good about their efforts. I wanted these to become his favorite canes. To sit prominently in the doorway and hear him say, over and over, "Diane brought these back for me from Down Under from her great adventure!" At least he couldn't return them.

In Wellington, Herr Most did something unthinkable: he took a day and a whole night off to stay with his wife. I simply could not imagine this stolid man in the throes of passion, not to mention smiling for an extended period of time. But I felt happy for him and hoped he was enjoying his conjugal visit.

Herr Stuhlemmer would also enjoy the marriage bed once again when his wife joined the ship in Panama City.

But by far the best bounty bestowed upon me in that Kiwi nation came in the form of a bulky packet delivered to the ship in Wellington by the Hamburg-Süd shipping agent.

"Fraulein Meyer," said Herr Stuhlemmer, "It seems someone at home is thinking of you." The bulk of the bulk came from a tape that my family had recorded for me. I sat in my cabin and listened to it over and over and over, until I thought the tape would break, just to hear their voices. My sister said hello and that she actually missed me (she was back from her Navajo visit). My mother hurriedly filled me in on all the local goings-on. She liked neither to be photographed nor recorded, and remained literally and figuratively in the background. Despite her beauty, grace, and intelligence, insecurity and low self-esteem dogged her. It was hard to shine next to my father's big, bright star. I ached to hear that in her voice, even as I reveled in the words. I rewound it and listened over and over again: "Love ya, my honey. Miss you."

And as always, my father stole the show. I knew the tape had been his idea, and I knew he knew how much it would mean to me. He read my spring semester grades that he'd opened when they came, even though they were addressed to me. He made jokes and gave weather and political updates. He and his longtime army buddy and great family friend, Joe, sang songs, and my father's notoriously awful voice made me laugh and laugh. I couldn't get enough of it, and it made me ache for home. I hoped they'd received the lengthy missives and pithy postcards I'd sent from each port, but I had no idea if any of them would reach home before I did. Most importantly, hearing their voices strengthened my resolve to stay on board and see the journey through to the end. I wanted them to feel proud of me. I had something to prove to them. I had something to prove to myself.

I did call them from Wellington to check in, but I could barely hear them over the very crackly, echo-y connection. They sounded so far away. They *were* so far away. I couldn't talk too long for fear that I would run up charges equivalent to the cost of the flight home I'd considered. I told them how much their correspondence meant to me, and about the potential delays and flight costs.

"My honey, come home if you want to." I could hear concern in my mother's voice.

"Klube," said my father, using my childhood nickname, "we love you. If you want to come home, you just let us know. Don't worry about the cost. But if you feel like you can hang in there, I bet the trip back will be a real adventure. . . ."

So there it was. A serving of unconditional love with a side of implicit pressure. My dad's *specialty du jour.*

I had already decided to stay because it meant something to me to know that I could. To test my strength, resolve, and resilience. Not for him. For me. I would never forgive myself if I cut and run in Auckland, no matter how challenging the crossing might prove. But now that he'd laid down the gauntlet, I picked it up without hesitation. "No, I'm staying. I love you guys. I'll be okay."

And Away We Go

July 23, 1979

36.8485 S, 174.7633 E

A big, bright-blue tug pulled up alongside to escort us out of the last port in the Southern Hemisphere. I'd never seen a blue tug before, and it looked funny against the red steel of the ship, bright and visible even though the sun hadn't quite risen. I had such mixed feelings about pulling away. I'd grown quite fond of this land Down Under, despite the relatively short time I'd spent ashore. The people I had met, to a one, treated me like family. They were as warm as the thick, dense wool covering the sheep that dotted the lush hillsides, with quick, easy senses of humor, and a ready willingness to lend a hand. Everything that the crew wasn't.

The tug let us go at the mouth of the harbor, and I waved at the few men on board, as if in saying goodbye to them, I could wish all the Aussies and Kiwis I'd met "G'day!"

The coastline receded as the sun came up from the direction in which we sailed, as if beckoning. Some of the lights in Auckland started to extinguish with the morning light. I imagined early risers putting on teakettles and going out to dewy front yards for the morning paper. A few vehicles moved on the roads—on the wrong side, I thought. Maybe delivering those papers, or milk. Do they still deliver milk here, I wondered, like they did in Howard Beach where I'd grown up? The glass bottles had indentation on each side, so our small hands could more easily grip them. The red plastic handles affixed around the bottles' necks allowed us to pick them up and out of the lidded metal box just outside the apartment door. We would leave two empty bottles inside on the night before a scheduled delivery, and the milkman would replace them with two of ice cold, fresh milk. Was it sheep's milk here? There was so much I still didn't know about this place, and I was already leaving. How differently my morning was starting out than theirs. The calm water of their harbor would give way to what were rumored to be some rough seas ahead.

Maybe some of the lorries, which now looked like matchboxes, were carrying the containers we'd dropped off. I watched from the deck until I couldn't see land any more, and tried not to think about when I'd next see it. But I knew that when I did, I'd be very close to home. I turned to go inside to report for duty.

Was I just imagining it, or did Herr Most actually look happy?

"Good morning. How was your visit with your wife? Is she okay?" I asked, hoping I wasn't prying too much.

"*Ya, ya,*" he said, clearly not interested in elaborating.

"When will you get to see her again?"

"Next time we are here. But after my next trip, I go off the ship in New York and fly home to Düsseldorf for a month. She will meet me there." And, to change the subject, he said, "And you, Fraulein, you have some friends on the ship now."

I had no idea what he was talking about. Did he think something had happened with some of the men?

"We have two new workaways. A married couple. Mark, he is from Scotland. And his wife, she is Barbara, from

Ireland. Maybe they are thirty. They stay below. They will work hard. Harder than you!" he pointed a finger at me. He said it in a half-joking, fully stinging way. I would happily have completed any assignment he gave me, but I also knew that he and everyone on board treated me with kid gloves because of Mr. Williams, the coordinator for their most important port, and because of the Herr Kapitän's edict. Having me on board at the whim of some American bigwig surely irritated everyone.

I had never asked for any favors or special treatment, but my simple presence was a favor, to Mr. Williams, and to a certain extent I inconvenienced everyone on board. I felt guilty for something I had neither intended nor could help. I sensed that he regretted what he'd said. He walked past me in the narrow galley and opened the dumbwaiter that had just arrived from below. Warm scents of dough, apple, and cinnamon wafted out. He carried a large pan back to the counter, silently cut a piece of fresh apple strudel, meant, I was sure, as part of the officers' breakfast—we never had pastry at breakfast downstairs—and put it on a plate. He handed it to me with a fork and napkin.

I mused curiously about the new people, and kept returning to her name. Barbara Barbara Barbara. My mother's name. As if she had come on board, through a surrogate, to shepherd me back across the Pacific. My mother had espoused mindfulness and the power of positive thought long before they became mainstream, while Eckhart Tolle was probably still sitting on the bench, homeless. She constantly "hodged" for things to happen or not happen, as the need arose. "Hodge" was her made-up word for focusing her thoughts and envisioning the desired outcome. She found parking spots for my father in the most impossible, unlikely locations in the heart of Manhattan. She channeled the Beach Boys' proverbial good vibrations to my sister and me whenever we needed a little extra help with a test or some other difficult situation. Later, after both of us married and had children of our own, she would "hodge" for them to make the golf team, recover from a tough bout of flu, or get into college. Things always worked out with her strong force behind us. I knew unequivocally that she was at home "hodging" hard for a safe and pleasant return journey. It was too coincidental that, out of all the names in the universe, the one female workaway we got was a Barbara. I knew my mom had a hand in this, and no matter how new age crazy that might sound to outsiders, only the believing mattered to me.

I was still mulling my amazing mother's powers when Herr Most broke my reverie: "You will continue to clean the bridge at 0700 hours. And Herr Rose or Stuhlemmer will contact you later to alert you, but now with the new workaways on board, you must switch to the port side lifeboat to balance things out. If we muster, go there."

My arrival must have caused imbalance. I was happy to do whatever I could to restore order and accommodate the crew, the new people, the ship, and the universe. This was such an insignificant adjustment, yet it made me wonder why I felt the need to meet their needs before mine. I resolved to take the time left in the journey to strike a balance of my own. Somewhere between Auckland and New York, I hoped to find a happy medium between obsessive people-pleaser and selfish bitch. Between worrying about others' welfare and neglecting my own. I hoped to find a bright spot amid all the dark blue. I'd made it this far; maybe this was the reason for me to complete the trip. So I looked at the crossing with new eyes, with new resolve. I had a mission.

I met Mark and Barbara at breakfast. They'd been assigned, or fell into, the outsiders' table with me and Karl, our lifeline to the crew.

Barbara, my mother's namesake, was solid, sturdy, and sensible. I could see that she would make a good sailor. Her halo of shoulder length, shag-cut hair frizzed permanently around her head in the humidity. She had a broad, freckled face and classic Irish-blue eyes. Hers was the third iteration of accented English that I'd heard on the trip, and her lilt lulled me as much as the others. Her straight, white, evenly spaced teeth surprised me for not being a typical British Isles mouthful of dental mishap. She smiled often, as if to show them off.

"It's so nice to meet you. We can't believe you've been on board since New York! You'll have to tell us everything we need to know about the ship," she said, sipping tea and then biting into a piece of buttered bread topped with liverwurst.

So I was the expert? The veteran? What an interesting tidal turn. I supposed it was true. As far a workaways went, I had seniority.

Mark was tall and gaunt, almost unhealthy looking. Sunken cheekbones protruded through the curtain of his full beard, and his hairline ran away from his forehead. Chunky knuckles punctuated the long continuous lines of his finger bones. He looked, I thought, as skeletal as Tim, and I wondered about drugs. When I'd see him working outside, shirtless, his ribs protruded for easy counting.

But there was nothing unhealthy about him when he opened his mouth. He spoke quickly in his Scottish brogue, which I could isolate absolutely only in contrast to Barbara's Irish when they were together. He spoke quickly and animatedly and seemed eager to plunge into ship life. This was clearly not someone on drugs. What a treat, I thought, to have so much pretty English to listen to all the way home.

"We've been traveling around before we go home to settle down—a honeymoon of sorts," he said. "This was by far the most economical and the most adventurous way to get from Down Under to the States." He thus reminded me that I was on an adventure, too. It slipped my mind occasionally. They'd met at university and married recently. His teeth were more UK standard issue, but he smiled sweetly nonetheless.

"Do you know what they will have you do for work?" I asked.

They both nodded. She said, "I'll assist Claudia serving meals and cleaning up the kitchen and the mess hall. And other general housekeeping duties, I guess."

Mark said, "I will paint the outside of the ship with the crew." Wow, I thought, hard work for both. My jobs seemed cushy in comparison, especially with Mark's. I knew that the crew painted the exterior continually, round trip on every trip, because the ocean was so hard on the heavy metal. I had seen the men, shirtless and tethered to the decks with harnesses and wires, like window washers high up on Manhattan buildings, except that those buildings were not moving forward at twenty-two knots through choppy water. At meals, their gloppy paint makeup told us on which area they'd worked that day: red-splattered skin for the sides of the ship, white-dotted for the superstructure, and black-streaked for some of the trim.

I didn't ask about Mark and Barbara's accommodations because I already knew from Herr Most that they lived below deck in hell, and I didn't want to flaunt my relative heaven. I never did invite them up to my cabin. We only met up at meals or in the common areas for gatherings.

It felt somehow reassuring to have them on board, but they kept to themselves and often retired early. They must have been exhausted by day's end, especially since we were losing an hour, and hence the corresponding sleep, every other day heading in this direction. I believed, as with small children when someone else's parent is around, that the crew would watch their behavior more with the two other sets of outsiders' eyes on them.

I didn't sleep well for the several following nights; the waves grew each day and conspired to make restful nights a distant memory. Maybe I ought to have worked as hard as they did, so I'd be more tired. In the evenings I watched the sun set in a spectrum of popcorn-butter-yellow, Creamsicle-orange, and cough-drop-red before I walked over to the other side of the ship to watch the moon rise. Sometimes the sky and clouds would conspire to hang a full rainbow over the whole spectacle. I'd only seen phenomena as breathtaking as this at sea. Witnessing this miracle every day at dusk—ever-changing in shape and intensity with the tides and temperatures—inspired as much awe the last time I saw it as the first. But for some reason, despite—or perhaps because of—the incomparable quiet in the middle of the ocean, which itself seemed quite restless with nary another thing in sight, I felt a vague sense of foreboding.

One night I fell asleep early, very early, at perhaps 2030. The time changes dogged me. At 2200 the fire alarm sounded, and not knowing if it were a real emergency or a drill, I scrambled to get out on deck—*port, port, port*, I kept reminding myself—in some presentable, or at least decent, state. There was no time to delay on a ship. If it were real and I got trapped up that high in the superstructure, I could only go up higher and then jump. I couldn't remember if I was supposed to go to the lifeboat or somewhere else in case of fire or fire drills, but only half awake, I couldn't think of where else to go, so I headed to my lifeboat. Fortunately it was only a drill, and fortunately my instincts were right, so I incurred neither wrath nor ridicule. The captain excused us with a nod after a few moments; we'd mustered in time. I zigged and zagged back to my cabin, in part because I was groggy, and in part because of the growing swell. At night, the waves were wraiths, their wrath invisible but strong.

Karl told me there were some nasty storms in the South Pacific on the route we'd normally take, so we were heading even further south to try to avoid them. I knew this meant more time en route. Maybe that haunted my sleep, too.

Cramps kept me from getting comfortable once I got back in bed, and I cursed myself for forgetting to buy more Midol or Feminax, the Aussie equivalent, before we'd set sail. I'd been too busy souvenir shopping. I chuckled at the thought of the likelihood that Herr Most stocked it in the Kanteen. Not much of a demand for it among this crowd, I imagined. I wished I had a beer.

I gathered a few of the new books around me—B.F. Skinner's *Science and Human Behavior*, Herman Wouk's *City Boy*, and J.R.R. Tolkien's *The Silmarillion*—trying to figure out which one would tire me out the most. In the end I dumped them all on the floor and pushed play on the eight-track player to let Genesis sing me to sleep.

But I thought Peter Gabriel got it wrong in *Cinema Show*. From where I sat, there was *way* more sea. . . .

———————

"Do you want to see below?" asked Karl at breakfast. I sat, glum, eyes at half-mast. I looked up at him. He never messed with me, but this sounded like some bad joke. Not him, too, I thought.

"For a tour. Have you gone below deck to see the ship's workings yet? She is an amazing machine. Would you like to see?"

"That would be so cool. I have to work this morning. Could we go after lunch?"

"*Ya, ya,*" he said, and I went to scour the interior walls. The air-conditioning never quite reached this core area enough to cool it, so the stultifying air kept me as wet as the walls I was washing. I couldn't wait to finish and get back to the chilled mess hall and the afternoon adventure.

Karl got me the smallest pair of coveralls he could find, but I still swam in the deep blue cotton. He laughed, but not in a mean way. "Still it is better to wear them. You can get very dirty down there." This would have sounded scary from anyone else in the crew. I'd been below with Herr Most before for beer, but that was only one level down. Karl, who had told me to wear sneakers instead of flip flops, led me down steep, narrow, textured metal stairs that turned back on themselves between each tier.

At home, the basements were the iciest part of everyone's home. Even in the throes of the hottest summer day, we'd go down into Sharon's dungeon to drink oversweet iced tea and talk for hours. Even though the upstairs had no air conditioning—they were just renting the house—each step down felt like a slow decent into a cool pool. Sharon, her mom, and her stepbrother had moved away to Texas after high school. After her stepfather died. Sharon died on her nineteenth birthday. She had stopped taking her epilepsy medication and had a seizure and drowned in her bathtub. In so little water, I thought now, in contrast to the vast amount that surrounded us. I felt as if I were drowning some days, in nothing but air.

But in the ship's bowels, the temperature rose as we descended; just the opposite of what I'd expected. I was Dante, following Virgil into the *Inferno*. I cannot remember if there were nine flights—one for each circle of hell—but there may as well have been. We just kept going lower and lower. By the time we could go no further, it felt vaguely like a way-off-the-strip casino. No windows, dimly lit, with lights flashing from some gauge at every turn. Clicking and growling and gurgling all around. I could easily imagine getting lost down here in every sense of the word. I could barely see, and I realized that I'd placed my hand gingerly on Karl's shoulder for guidance and security. Thankfully the narrow walkways had railings on both sides. I held on tightly, glad for the barrier between me and the labyrinth of asbestos-covered pipes, and mélange of metal I might fall into with one misstep.

"Don't touch anything," shouted Karl. It was hard to hear him over the mechanical chorus.

Don't worry, I thought, walking through this whale's abdomen. I had no intention of poking and prodding the metal innards.

"These are the sea water tanks," he said, as he pointed at the immense containers that pulled in salty water and desalinated it for cooking, showers, and, I realized, drinking.

Further aft we approached two even larger dark steel cubes. Together they might have filled one of the containers on deck. "And these are the boilers. The steam from these powers the turbines," he said, pointing them out as we continued on slowly, as if he were a host at some surreal car show, highlighting the latest model's fancy features.

The turbines roared at us like two trapped lions, their clawing and thrashing channeling energy to power the massive vessel. I could barely hear him now, and the heat, especially swathed shoulder to ankle in thick blue, was almost unbearable. We walked even further back—I could just make out a monolithic mass that must have been the stern. It was so dark that the devil red down here appeared nearly ebony. The few light bulbs strewn too far apart overhead barely impacted more than a small space around them. It felt as if the turbine's reverberation were Roger Waters himself welcoming me to this machine. I'd played that Pink Floyd eight-track in my cabin on an endless loop, too.

"And," he paused for effect, "the turbines turn the propeller." He pointed at and past the stern, where the propeller turned eternally, inching us along. I had no context for the scale of this mechanism. We couldn't see the propeller, but I felt it, and silently thanked this nearly two-story, hundred-ton hunk of metal fan that moved me ever closer to home.

As we wound our way back up, the deafening roar tamed to a steady hiss. "She is a beauty, *ya?*" asked Karl, clearly proud, as we emerged into the refreshing air of the superstructure.

"*Ya, ya*," I said. "She is impressive."

I could not peel the coveralls off quickly enough. I handed them to him, a little self-conscious because they were dripping. I was as drenched as if I'd just jumped into the small deck pool, which sounded very appealing at that moment. He laughed again. "*Ya*, it is so hot down there, no?" I nodded, with a new respect for the poor souls whose jobs kept them below deck for their entire shift: engineers, electricians, machinists—and awe for what this ship did every day with serene efficiency.

She showed off now by bringing us to full speed since we were free of pesky landmasses. The forward motion minimized the impact of the immense sea surge, and it felt like progress.

Two crewmembers, Walter and Franz the baker, celebrated their birthdays the next day, so that meant two beers per person. A sense of relief washed over the ship, similar to the one I'd noticed after we got through the canal, but even more palpable. We had successfully delivered and taken on all the required cargo without incident. The captain's mood impacted everyone, and his relief at the accomplishment infused us all with light.

"So come to the party tonight for Walter and Franz," said Karl at the breakfast table. "They are both Austrian. They will sing and dance. It will be fun. Right here, in the mess hall." He smoothed his moustache and smiled.

I hadn't had great luck with theses large gatherings. The men tended to get very drunk and act very badly. But the festive mood of the moment swayed me to try again. I had no idea why Austrian equaled fun in Karl's mind, but I took his word for it. I hadn't even realized until then that, aside from Tim and the phantom laundry man, anyone on board hailed from anywhere but Germany.

The party started as they all did, with a big group of tired men guzzling beer, and I sitting tucked into a corner, clutching my Holsten bottle like a life preserver, and peeking around furtively trying to blend into the Formica.

Ingo slid in next to me, his apron-less belly kissing the table, and nudged me with his elbow. "*Wie geht es*, Fraulein Meyer? Nice to see you leave your cabin."

We clinked bottles.

"*Prost!*" he said.

Our eyes popped, and we looked at each other askance when Herr Most took the seat across from us. I couldn't hold Ingo's gaze for fear we'd gasp, betraying our incredulity, and scare him away. He was the most diurnal creature on board. After dinner service, when he'd finished tidying the galley or closing the Kanteen, he would retreat to the comfort of his cabin and rarely emerged after sundown. Like a reverse vampire, threatened by the dark.

"*Abend.*" He nodded curtly to me and then to Ingo.

"*Abend,*" Ingo returned.

"Hi, Herr Most!" I said. He lowered his eyes at my enthusiasm. His hands covered a small cellophane-wrapped bundle which he slid toward me. "My wife gave me a large box of chocolates for the crossing. I don't know why. She knows I do not eat so many sweets. It is okay for you—you don't worry so much about getting fat." He smiled, but kindly.

Ingo jiggled with laughter as I thanked Herr Most, and said, "You must share? *Ya?*"

"*Nein, nein,* they are not for you, Ingo. You do not need any," and he pointed to the place where the table dented Ingo's soft midsection.

We all laughed as I unwrapped the treats and put one in front of Ingo. So Herr Most did not reserve his ribbing for me alone? I felt less singled out and more included, as Ingo pushed the dark square back to me. "*Nein,* your boss is right. It is for you."

The cacao coated my tongue in dark bliss. Maybe the connubial visit had softened Herr Most's sharp edges just a bit.

The birthday boys rose from the gathered revelers as

everyone lifted their Holstens in a toast. A few gave individual, roast-like tributes in German that I completely missed, but the laughter told me that they were in good humor. Someone pushed play on a portable cassette machine, and festive Austrian music saturated the stuffy air.

Ingo said, "*Ya, ya*, here we go."

The birthday boys started a round of intricate call-and-response clapping and toe tapping, and soon the room ignited with the clamor of hands coming together and feet pounding the floor. They swayed in time with the music, and then a few got up to dance. We could resist the tug of the tunes no more than the oceans can escape the pull of the moon. Soon we were all on our feet, heels slapping down hard and knees rising up high, hands hoisted overhead, with the Austrians, whooping and hollering words to the traditional songs. It was a regular whatever the word is in German for a hootenanny. It felt cathartic—a communal primal scream.

The energized, sweaty well-wishers then watched as the two honorees stood on chairs and began yodeling back and forth. Who knew that these rough, gruff men could produce such dulcet and dainty sounds? Robust, yes, but as clear and crisp as Alpine water.

It was far and away the most exhilarating, truly fun night I'd had on board. No one leered or lunged at me, and as I let my hair and my guard down, they welcomed me in. My feelings about this crew swayed as often as the ship in the swell. Sometimes, especially after something like this, I felt part of an exclusive, foreign fraternity. But more often I felt like the frat pledge who had been hazed and then rejected.

That evening, and for many of the next few, my cabin phone started to ring again in the middle of the night. Sometimes the caller identified himself: "Fraulein Meyer, it is Herr Stuhlemmer here." He slurred his last name more than I had when

I'd first tried to pronounce it. "I hope I did not wake you." He tittered a little bit, and I thought, *What the fuck, it's 0238, of course you woke me!*

"I had to tell you," he continued, "I love you. You are so beautiful. I wish I had never married my wife, ugh, so boring, so fat." I thought of poor Frau Stuhlemmer who was flying half way across the globe to Panama City to spend some time with her Odysseus.

"Herr Stuhlemmer. You've been drinking Schnapps." I hung up.

Sometimes the caller remained anonymous, but the pitch was predictable: "Fraulein Meyer, we have a party now, why don't you come down to the Kino lounge." Half awake, I couldn't distinguish voices and didn't care enough to try. I didn't even answer before replacing the receiver on the cradle. It upset me enough, though, that I always had a tough time falling back asleep, and the morning wakeup call to clean the bridge found me groggy and irritable.

I mentioned the unwelcome intrusions to Herr Most one morning after he handed me a warm slice of sweet raisin bread. He looked down at his, slathered in the butter that seeped into whatever spaces there were between the raisins and warm, soft dough. (Where did this man put all his calories? He had no discernable fat reserves. His engine revved high.) He said nothing.

Later that afternoon, while I was carefully cutting out images of tikis from the New Zealand museum pamphlets for my collage, there was a soft knock at my door.

I cracked it open to see Herr Kapitän Beucking himself standing there, head tilted down toward a small black case he held in front of him in both hands. Only his eyes looked up at me from under his well-endowed eyebrows.

"*Guten Tag*, Fraulein Meyer. I am sorry for the disturbance."

I immediately sensed that it was somehow I who ought to be apologizing, but I wasn't sure exactly why.

"I understand that some of the men . . . uh . . . call you in the night. I imagine when they drink. I am sorry for this. May I?"

"Oh, of course, come in." I still wasn't sure where we were going with this.

"It is better that you do not have the phone. You do not really need it. I am right next door. I do not want you disturbed any longer when you sleep." I stepped back and surreptitiously scanned the cabin. Fortunately it was clean except for the craft detritus on the table.

Within moments he had removed the phone from the wall and disconnected the wires. He held the whole apparatus in his hand, bowed his head slightly, and said, "I hope this helps. I trust the men will disturb you no more."

"Thank you, *danke*," I said as I shut the door behind him. He'd been in the cabin for less than five minutes.

I sat back down at my table to work on the collage, but couldn't focus on cutting or gluing. Herr Most must have told the captain as soon as I left the galley, clutching my warm raisin bread.

The captain wouldn't have let any of the men come remove the telephone because he couldn't know the perpetrators' identities. I was especially relieved not to have mentioned the radio officer by name. But I imagined the captain sternly rebuking the men: "It has come to my attention. . . ." Would he convene a meeting? Send a memo on the thin onion paper that they used for morning news? Or relegate the reprimand to each group's supervisor?

I shuddered to think of the retribution I might incur for having opened my mouth, and realized that while this meant that my nights would be more restful, I also no longer had any way of calling out once I was in. It was a mixed blessing with even less connection to the world outside than before. Herr Kapitän had mentioned his proximity as an assurance of safety. He'd told them not to bother me before I boarded, but beer made some of them forget. He'd remind them, though, I felt sure.

Fuck 'em, I thought. Let them be pissed at me because I stood up for myself. Shame on them. Let them not talk to me. I'm on my way home. And I'd become quite adept at waking up on my own with the little red leather-wrapped Bulova travel alarm. Who needed wakeup calls? And whom did I need to call anyway? Once ensconced in my personal container, with the door locked, I was perfectly safe. I lived next door to the captain, as he himself pointed out. No one was going to fuck with me up here.

At schmoke time I started to tell Herr Most what the captain had done. He cut me off with a wave of his wiry hand. "*Ya, ya*, I know. Now they cannot bother you."

"*Danke*," I said, but he waved the thanks away, too, and handed me a warm éclair.

Herr Rose pointed out Pitcairn Island one morning while I was cleaning the bridge. Very beautiful, and all I could think about was the equally beautiful Marlon Brando and his breadfruit in *Mutiny on the Bounty*. And then about mutiny. I'd like to mutiny, but against and with whom? They'd make me walk the real plank into the infinite azure.

"We will pass Easter Island today, too," he said, snapping me out of my rebellious reverie. "You know, where the enormous moai statues are."

"I know the statues, but didn't know what they were called. Will we be able to see them?"

"*Nein, nein*, we will not sail so close. You can see only the island, like we can see Pitcairn."

It was July 29. I really wanted to ask him how long it meant, then, before we got back to the canal, but I kept my impatience in check. Just enjoy the journey, I reminded myself. And anyway, even if he told me, it was bound to change ten times before we got there. These waves often dashed false hopes.

When I checked in with Herr Most in the galley, he handed me some confectionary creation covered in baby pink frosting. Clearly he counted himself part of the ship-wide conspiracy to maintain my heft. I don't remember how it began—maybe I'd mentioned *Slaughterhouse Five* to him—he always asked about what I was reading—but we started talking about World War II.

Herr Most, like Billy Pilgrim, and like Kurt Vonnegut, Jr. himself, had experienced and survived the firebombing in Dresden during the war. Herr Most hailed from that ill-fated city. He was twelve when it happened, which would make him forty-six, the same age as my father. I could hardly believe this because he seemed so wizened, so withered, and so much older in many ways, despite his stamina and strength. He spoke quietly of the raging fires, of the leveled rubble, of the loss of so many friends. No wonder, I thought, he too had become a pilgrim of sorts. No wonder he chose to spend his life surrounded by cool and extinguishing water. No wonder he was such a curmudgeon sometimes.

I had only heard about "the war" from my parents and grandparents, none of whom had experienced it firsthand. They recounted the relatively mild hardships: shortages of butter and sugar, managing ration stamps. I'd heard joyous tales of V-E and V-J Day celebrations, from the victor's viewpoint.

I'd interviewed an American veteran who had participated in the Battle of the Bulge for a ninth-grade world history paper.

"I want to write about World War II," I'd told my teacher, Mr. Warren.

"You need to narrow your topic, Diane, you're not writing a book," he cautioned. So I settled on that pivotal, arduous Ardennes Forest engagement. We needed a "primary source," so I asked the father of a boy I babysat if I could interview him. Looking back, I realized he resembled Herr Most with an extra foot of height and a head of brown, not silver, hair.

"I have nothing to say about it," he told me, more melancholy than dismissive. "It was cold and dark and scary as hell." You can put that in your paper." I did. That was good enough for me; the quote counted as the primary source I needed. I would never forget his words. I got an A+ on the project, and a hint—a whisper—of what this war thing might have been like for real, beyond the pages of my ninth-grade term paper, beyond the pages of Jerzy Kosinski, Ernest Hemingway, Stephen Crane, and all the authors who tried to make noncombatants understand their own private hell.

Here, again, very close in this narrow stainless steel galley, I got a whiff of what it smelled like to live in a city that blazed, then sizzled, then smoldered. What it felt like to stumble along *Strasses* he'd easily skipped down before, negotiating an obstacle course of strewn and tangled stones and rafters, and, in some cases, charred limbs. What it sounded like to listen to the shrieks of people melting, no matter how tightly his mother covered his ears and pressed his head to her breast; to hear the victims' wailing and then whimpering bathe him in a cascade as relentless and chilling as the raging of a mountain stream in spring.

No wonder he, like my neighbor back in Westfield, did not want to talk about it too much. I told him I hadn't meant to pry.

"*Nein, nein*, it is okay. Sometimes we have to talk about it. It was a long time ago." He looked somewhere past me now, not down at his hands like usual.

"Did you ever read the book?" I asked.

"*Nein, nein*. I don't need to read it. I lived it."

Fair enough, I thought, fair enough.

———

That night, the engine's rumble, and all its attending rattling, banging, and vibrating, abruptly stopped. The noise and motion of the colossal vessel, which was, at the beginning of the trip, distracting, had become more of a soothing meditation mantra. So when everything suddenly ceased, I sat up and took notice. Literally. The only thing more pronounced than the silence and my heartbeat was the ship's undulating in the monstrous mid-South Pacific waves. The swell had reduced the big red behemoth to a small buoy, now unimpeded by forward motion. It was nauseating.

The pitch black told me there was no moon that night, or that clouds covered it. The sound of distant footfalls and muffled but urgent-sounding snippets of conversation slipped under my doorway. I fumbled for the clock. 0200. What the hell?

According to the original schedule, not worth the red paper it was printed on, we should have been nearing New York by now. We, or more precisely, I, could not endure any more delays. At this rate, I had no faith that I would make it back in time for the beginning of my junior year.

I should have flown home from Auckland, I thought, as the strong oscillations reeled me up to the head of the bed and then tossed me back down to the foot like a fish caught on a line. Maybe I'd fly home from Panama, if we ever got there. But I really didn't want this summer sojourn to cost any more than it already had, and I still had this romantic notion that, like Odysseus, I should complete the journey. I barely slept, tossed by waves and tormented by thoughts.

"*Guten Morgen,*" said Ingo mischievously. He couldn't wait to pounce. "Not much sleep, eh? The sea is very rough. She tosses us like a salad." His culinary humor amused him. "We are stuck, *ya*? The boiler is shut down." He grinned from ear to ear, and rubbed the white cotton covering his belly as he awaited my reaction. Claudia brought me a carafe of tea along with bread, butter, and an array of pink fatty coagulated cold cuts. "*Morgen,*" I directed my greeting to her. The meat turned my stomach.

Shit. Shit shit shit! I could absolutely not let Ingo see my distress, but I was having a hard time suppressing it. "Oh, broken boiler? Interesting."

"*Ya, ya.* It could be days."

Shit.

On a normal day, we could always see the sea and sky simultaneously out of the portholes in the downstairs mess as we listed languidly from side to side during steady forward motion. Now, standing still, the downward starboard tilt was so steep that we seemed submerged. All I could see was an agitated sea out the porthole, as if I were inside a washing machine, on spin cycle, looking out. The upswing tilted us so high that it showed only dark flannel gray sky. The ship had become a carnival pendulum ride that was swinging us a full seventy degrees side to side in the middle of the ocean. With each slow, deep roll, the competing hues changed places: water, sky, water, sky. This was the steepest pitch we'd encountered, so I tried hard to fix my gaze and get something in my stomach so the motion didn't make me ill. I wouldn't let the men see me succumb to the lurching. To the ocean that bounced this steel whale up and down like two teenagers on a seesaw.

"*Ya, ya*. The engineers work on it now. They say maybe they will put you inside the boiler to have a look around. You are so small! You will fit!" Wider grin, still.

Surely he jests, I thought. He won. He got me.

"Ingo, stop it! No one wants to put me in the boiler, and anyway. . . ." but I stopped to follow his eyes to the door of the mess to see Chief Engineer Schnoor standing there in his dress whites. Which meant he was on duty. Officers rarely came down to the crew slums themselves. Usually they sent a messenger. He was an imposing figure in the doorway, his outstretched hand resting on the top of its frame. Ingo stopped grinning and stood taller. Claudia slipped into the galley.

"Fraulein Meyer. *Guten Morgen*. Will you come with me for a moment, *bitte*?" I looked back at Ingo who now looked down. He would not meet my gaze. Claudia had disappeared. "*Schnell, bitte*."

"*Ya, ya*, right away." I left the hot tea and cold meat where it was. I had no appetite any more anyway.

We walked out together, he briskly, marching really. I tentatively, sort of chasing him. "*Wie geht es*, Fraulein Meyer?" He asked, but before I could answer, he added, "You notice we have stopped. We have some issue with the boiler. You had a tour of the ship mechanics, *ya*, so you saw it, I understand? I was nodding, but wondering where this was going. "And you said you might be willing to go inside during the cleaning?" Now I stopped nodding my head and started shaking it vaguely instead.

"Well, um, yeah, no, I—"

"We need to determine what is happening inside, and perhaps if you are still interested. . . ."

We followed the same path down as Karl and I had taken, but this time I felt more like Gretel being led to the oven than an intrepid adventurer. And as the lone Jew on a German container ship, the import of this image was not lost on me. I had no idea if anyone on board knew my religious background, nor had I ever discussed it. I was hardly devout, or even practicing, but I do come from a lineage of Orthodox practitioners, including some rabbis and cantors. Our family on both sides lost distant relatives to the Holocaust. I knew intellectually that this was a completely inane rabbit hole to think myself down, but I began to feel dizzy glancing over the precipice. Why didn't they ask Ana? She was even smaller than I. She'd fit really well.

"Of course if you are uncomfortable. . . ." He continued to talk although I'd missed the last moment or two of what he'd said. *Nauseous* was closer than *uncomfortable* to what I was feeling.

"You can just have a look in," he said, as if encouraging me to peek in at a basket of sleeping newborn kittens. He pointed at a stepladder that would get me to the window-sized opening, about six feet up. "Be careful, even at rest it retains quite a lot of heat."

No shit. I could feel the warmth radiating off it. With one foot on the first rung, I hoisted myself up and looked into a cavernous abyss. The heat alone could crisp my face like the sun at the equator. Did the mottled, whitish walls glow a little orange, or did I imagine that because it felt like hell?

I imagined Herr Schnoor, like the witch in the woods, pushing me in, closing the heavy wrought iron door, and latching it behind me. So this is how they planned to get rid of me. What would they tell my parents? That I fell in during a valiant attempt to assist them? I could not breathe the red hot air into my lungs, and I felt as if I were drowning, but not in cool blue water.

I wanted to help. I wanted to be helpful. To be the hero. To have them like me because I helped them solve the boiler problem and got us moving again. But then, I thought, forget it. Pleasing them was no longer my priority. I was my priority. I was scared, and even if it meant further delay, I decidedly did not want to descend into that infernal container. Whoever went in to clean it could go in again to figure out what was wrong with it. I may not have had a clue what I was signing up for when I came on board, but it was decidedly not this.

"Herr Engineer Schnoor, I am so sorry, but I feel really uncomfortable right now. I'm having a hard time taking a deep breath. I think maybe it'd be better if someone who knows the boiler better—"

"*Ya, ya*, Fraulein it is not a problem at all. Really I just thought if you were curious. We will have one of the machinists take a look. Thank you for coming down. *Guten Morgen*."

I turned and went back to air conditioned oxygen as quickly as I could without looking as if I were fleeing a crime scene. It occurred to me that I might have misjudged his intentions. Maybe he'd heard I'd been down there poking around with Karl and thought I'd find an inside tour of the boiler interesting. That they'd be doing me a favor—adding some spice to my otherwise pretty mundane days. Or maybe they just thought it would be easier to maneuver a smaller person around in there than some of the heftier engineers. I didn't know, and I didn't really care.

I didn't want to go into the boiler. For any reason. And I didn't care what they thought.

Lost and found. Lost in space. Lost at sea. At a loss. I once was lost but now I'm found.

When I was six, my mother lost me. Not on purpose. She was struggling to juggle shopping bags and a three-year-old Suzanne in her arms and held tenuously onto my hand. We jostled along in a crowded rush-hour bus en route home from a shopping trip in Jamaica, Queens. She strained and swayed to keep her balance and a hold of everything, a slice of meat sandwiched between sweaty bodies. At six, I could hardly see her, or anything except belt lines and bellies smushing up against me. Her hand was my lifeline.

Someone pulled the overhead cord to signal that they wanted to get off—I heard the dull *ding-ding*, and the bus jolted to a stop. My grip on her loosened, leaving me afloat in a sea of strangers. What I knew a few scary moments later was that it was not *a* stop, but *our* stop, and she, the packages, and my sister had disembarked. I was lost.

"Mommy! I want my mommy!" I wailed as the black rubber-lined accordion bus doors closed, the diesel engine revved, and we pulled away.

Passengers must have seen her screaming as she dropped the parcels and my sister to the ground to frantically wave her arms: the reverse image of Zhivago frantically trying to get Lara's attention from the bus as she walks calmly along the street and out of his life for the third and final time. In that scene, Zhivago did not succeed and lost Lara forever. The passengers surrounding me joined ranks and got the driver to stop so we could be reunited for a happier ending.

Everyone always told me I had a good sense of direction. That I know where I am in space. "It must be a gift," they say, like Zhivago's brother Yevgrav tells his niece Tanya, Lara and Zhivago's child, about her balalaika skills. She too lost her mommy's hand in a crowd.

This, I thought, is not so much a gift as a skill built through sheer determination never to be lost again. To hold tight to my lifelines, but to know where I am if they slip away again.

I could have asked—and often did ask—the duty officer for our exact longitude and latitude. But that information, while accurate and precise, would have given me no more orientation than the very few and far between and completely unfamiliar landmarks that marked our progress. I needed to develop an inner compass to find my own center of gravity, regardless of how the crew treated me, how unfamiliar my surroundings, or how the ship tried to throw me off balance.

By August 1 the storms had passed. Not surprisingly, the crew had managed to repair the boiler without me. We had crossed the International Date Line for the second time and had two Mondays, complete with repeat Monday menus. It was funny, this manipulation of time. I wished someone could teletransport me home, manipulating my molecules like they did the clocks, but at least the advent of August made the return home more tangible, as if I could see it in the distance through a spyglass. Barring any more unforeseen delays, we would dock in New York this month. I redoubled my efforts to trim down a bit. Like the meals, the schmoke-time treats reappeared in a predictable loop, so I'd sampled most of them by now. Unless something really special appeared (anything chocolate), I resisted. Also, I decided to tackle the other side of the equation with actual exercise. The push-ups, sit ups, and running in place in my cabin had no impact, and anyway, after Wolf almost threw me overboard, I was loath to do anything but tiptoe around my cabin.

Voyeurs be damned, I donned a pair of gym shorts and my rattiest T-shirt and slid into the only-marginally-bigger-than-bathtub-sized pool most days to do laps. Lots of them because of the truncated length. I still felt self-conscious and tried to strategically select times when no one else was apt to be out there, but the water (salty) and the motion (my own, not the ship's) felt invigorating. Propelling myself for a change, rather than surrendering to Neptune's whims, empowered me.

After Kino one night, Barbara and I sat out on deck and talked. It was nice to have the company, in English, of another woman. I was disappointed that we hadn't spent more time together. When she came on board in New Zealand, I'd had fantasies of having a new best friend, a lifeline to save me from drowning in this pool of testosterone; but she was busy, tired, and married. She came alone that night. Mark was sleeping already, but she couldn't. The movie was the 1974 *White Dawn*, in English with German subtitles, about three sailors stranded in the Canadian Arctic with the indigenous Inuit. We mused at the irony of showing a ship full of sailors a movie about shipwrecked men.

"I could so relate to that," I said to her, laughing, watching the moon paint the crests of the frothy waves white and tint the tips of the flying fishes' wings silver. "Except it's hot and humid here, and the natives have not been nearly as welcoming," I added.

"I expect not," she said. "How has it been for you on board, all alone?"

I gave this question more thought than I'd allowed myself now that I had to answer it out loud.

"Oh, it's been okay. Really it hasn't been bad. But it hasn't been good either." She nodded. "At first I was a novelty to the crew, but one protected from the top down. Once they realized that I would not sleep with any of them, they shut me out for a while, but now they just treat me with indifference. Karl is great. Ingo likes to give me a hard time, but he's harmless. Some of the crew can be rough, but the officers are generally kind."

"But aren't you lonely? Don't you miss home?"

My eyes welled up before I could answer. I just nodded. Again, having to say it out loud was like bringing a submerged submarine to the surface. It kind of sucked the air out of me when it crested.

"Yes. A lot. I'm not sure what I expected or what I thought I was getting into, but it's been quite an odyssey. I'm really glad I did it, but I can't wait to get home. Back to my family. Back to my friends. Back to school."

She touched my shoulder. "I get it. It's different for me; I have Mark. And there are things at home I needed some distance from."

"Do you mind if I ask what?"

"No, of course not. At University, before I met Mark— we both read philosophy—a boyfriend raped me. I fell into Mark after that. He enveloped me, and I felt protected and safe. But I'm not quite sure I'm over it. Do you ever get over something like that?"

"I'm so sorry. I cannot imagine. I think Bruno is rough with Claudia—don't tell anyone I told you that. What is it with guys that they think they have that right?" I asked.

"Oh, it's not all of them. Mark is a love. You just need to make good choices. But that's true in general, isn't it?" I nodded, and wanted to ask how to make those good choices, but she went on.

"And my brother, my baby brother, has had a very difficult time. My parents finally had to put him in a mental hospital. I've no idea what will happen with him, and it's agonizing not to be able to help him."

I could do nothing but nod again. I had no context for any of this. My problems seemed now like splashes in a backyard kiddie pool compared to the tidal wave she was facing.

"The flying fish mean we're getting closer to the canal. I first saw them when we came through to the Pacific," I explained.

"I've never seen anything like them before. What fantastical prehistoric creatures!"

I laughed into the deep blue night. "Yes, I do feel like I'm in some Neanderthal fantasy sometimes." Now she laughed. "You know we crossed the equator today," I added.

"No! No one said anything. We flew over it last time. I did see the Galapagos. I wish I'd known."

I have to admit I felt a little tiny bit smug at having been awarded the equator-passer certificate, even if it had come with an ulterior motive.

"Yeah, we're back in the Northern Hemisphere. The toilets' flush and the sinks' drain have reversed. The Southern Cross has disappeared, and the North Star has taken its place," I said, pointing up at the inky sky.

"Yes, everything is right in the world," she answered.

Homeward Bound

August 6, 1979

8.9824N, 79.5199W

We could see the coast of Panama on August 6. Despite the bad weather and rough seas, the return crossing still took five days less because New Zealand was further east than Australia. I no longer felt so much that the ship propelled us, as I did on the way down, as that the land—and home—pulled at us now. As if my parents had pushed me out of the nest, a gray downy cygnet to test my mettle. Now I sailed back proudly as a full-grown swan, able to survive on my own.

We anchored just outside the canal to await permission to enter amid an armada of other attending ships. A coterie of dolphins greeted us as we glided into place. As before, the glimpse of land delighted me, as did the sweet sound of the Spanish-speaking DJ playing American disco songs on the radio that could finally receive a signal. Barbara and I finished work by lunchtime, so after our meal we sat by the pool protected by the deck overhang. I knew the near-equatorial UV rays could penetrate even the fog that had followed us for most of the Pacific crossing, and I'd learned my lesson about sun exposure on the way down. Her fair Irish skin avoided solar rays like Dracula. We watched Ana splash and giggle in an off-white crocheted bikini that barely covered . . . well . . . anything.

As she frolicked, I kept trying to make out the coast through the fairly heavy cloud cover. I felt a completely different type of anticipation than I'd had coming in the other direction. Then, I had no idea what to expect, from either the canal or the wild blue thereafter. But now I knew. I pointed out each of the different ships to Barbara, "The one that looks like a tank transports cars, and the smaller one next to it carries pallets of cargo." I explained the complex canal lock procedures and told her what we'd see on the journey through. We'd come to the river portion first from this direction, and the wider, more lake-like part toward Colón, at the Caribbean side. I knew with more certainty how long it would take to get home from here. We would head straight up to Halifax, and then Boston, before New York, but we'd be in and out of port quickly, so there was less opportunity for delay. I could count the remaining days on my fingers.

I could easily and much less expensively get off in either port and travel home by plane or train if I wanted to, but curtailing the remaining few days after Halifax or Boston seemed pointless. No, I could see the finish line from here. I would cross it, like I had the equator and the International Date Line.

The crew knew I would disembark soon and became suddenly friendlier. Many had begun to ask what I thought of them, the trip, and the ship. Some gave me their addresses at home, and told me to let them know if I ever travelled to Germany.

"And next Friday, we make 'schnitzel a la Meyer' in your honor, Fraulein!" announced Ingo. A dish named for me? Like a regular celebrity? Ingo couldn't know about our miniature schnauzer named after the iconic German dish. Suzanne could barely pronounce her name at age five when she debuted, slipping and sliding down the hallway in her own pee, tail wagging and nails clicking and clacking, yapping a high-pitched greeting. "Shit-zel!" she'd say instead, making us all laugh. At eight, I could just make my mouth cooperate enough to get her name right. I missed that dog as much as any human, so Ingo's culinary tribute took on even more meaning.

The attention surprised me; I didn't think anyone even noticed me anymore. I'd become more like one of the containers, just something to load in one port and unload in another, just so much cargo. But it seemed as if I'd acquitted myself well enough to dispel any notion of the rich, spoiled, preppy, college kid they thought they'd seen when I boarded.

I felt a little wistful myself, and thought about inviting some of them up to the lake house when we docked in New York. I wondered if my mother would welcome a bunch of sailors on shore leave for a typical American BBQ, but quickly dismissed the idea as absurd. For a number of reasons.

It was no less fascinating to watch the canal crossing in reverse the next day, as we entered the intricate lock system that would lift us from sea level up to the level of the canal so we could sail back through.

The Colón agent had only a few letters from friends for me, as well as Frau Stuhlemmer for the radio officer. She was a jolly, red-cheeked dumpling of a woman. I could not help but envision her wearing a traditional German dirndl, mastering the two-fisted carry of multiple tankards of frothy ale to a rollicking Oktoberfest *biergarten* crowd. Herr Stuhlemmer could hardly look me in the eyes while she was on board. When he did, it was with a combination of embarrassment and veiled threat. *Don't you dare tell her about my drunken dialing and confessions of love. Please!* I imagined him thinking. He needn't have worried. I had empathy for her and had no intention of ruining their time together.

On the other end, the Panama City representative brought letters and another tape from home. I was happy to escape the heavy heat and head back to my cool cabin to devour them.

Once through the canal, heading north, the sirens of home seemed to call from everywhere. I listened to a soap opera from Cuba, a Bahamian radio station, and finally a pop AM station from Miami. It was as if, with gentle tethers, they were gently guiding me, reeling me back in.

"We stop in Charleston tomorrow," said Herr Most, offering up some fresh-baked apple strudel that I refused.

"I thought Halifax first?" I asked. Three more ports instead of two. I tried to recalculate the estimated time of arrival again in my head, but gave up, my shoulders relaxing. It doesn't matter, I thought. Soon enough. It wasn't weeks. It was days.

"*Ya*, but these next three ports, they are very short stops, sometimes not even overnight, depending on when we get in. We drop some containers from Australia and New Zealand, but we do not load anything again until New York. You can see if you look aft; the ship, she is mostly empty. She has lost some weight, like you. Not eating treats anymore?"

I had noticed the lightness. Without the containers stacked high behind the superstructure, the sunlight could make its way through the previously blocked portholes. The ship lifted slightly above the surface, having shed some of her burden.

"And *real* workaways disembark at the first port, so your new friends will go off in Charleston."

I didn't know that either. Barbara, not that I'd seen much of her, hadn't mentioned it. I had no right to expect that she'd tell me, but I felt a little disappointed nevertheless.

"We passed through the Bermuda Triangle today. No disasters. Smooth sailing from here, Fraulein Meyer," he said.

This was as chatty as he'd ever been. Maybe he felt lighter, too, knowing he would soon be relieved of his own burden. Me.

The Atlantic looked more welcoming, too. We had left the murky, churning chop behind us on the other side of the canal. The wind pushed us from behind now, rather than fighting us from in front as it had all the way up from New Zealand. The water was clear, crisp blue and provided no resistance. The bow sliced through the surface like a glazier scoring a mirror.

My grandfather David, my mother's father, was a glazier. I remember watching him wield a wheel cutter in his workshop in Union City, New Jersey. Careful, serious, and concentrating hard, he slid the blade over the reflective surface. We could not make a sound lest he slip and break the fragile glass, or cut himself. The ship now seemed as focused, determined, and skilled as he, as she made one long, smooth straight slice, barely creating any waves. None of the strain and struggle that seemed to accompany every mile in the South Pacific. The aft wake was calmer, too, and the sun shone brightly. The extra stop in South Carolina could hardly do anything to dampen my mood on such a day.

That night Barbara and Mark treated everyone on board to a beer. We toasted at dinner, and I asked if I could take a photograph of them.

She shook her head adamantly. "No. No photos. You don't need a photo to remember us. You know, in some cultures they believe that you steal someone's soul when you take their picture."

Her reaction surprised me, but I respected it. We docked before dawn in Charleston, and they had disembarked before breakfast, without saying goodbye, almost as if they'd never been there. I wondered why they had left so early—maybe the ship required it—and if I'd just imagined them to help palliate the pain of the Pacific passage. My imaginary friends.

After breakfast I went into the port office to make a call and ask how to get in to town, although Herr Most had cautioned me not to take too much time because we'd probably leave before dinner.

Manning—and I do mean manning—the office were three delicious southern gentlemen. Hot as steamy grits, each sat at a military-issue-looking steel desk on stiff metal chairs resting on a gray linoleum floor. The grim surroundings made their sweet smiles and sky-blue eyes even more alluring. Their crisp navy uniforms said military too, but I knew they were Port Authority. I almost had to pinch myself when one, who was a blond version of the Middlebury football player with a crush on me, said, "How can we be of service, ma'am?" His drawl, butter dripping on the grits, was one more ingredient in the mélange of accents I'd heard during the trip. They were the first Americans I'd seen in over two months, and what fine specimens to welcome me back they were!

"I need to make a call home, and then I'd like to figure out how to get into town to do some sightseeing before the ship leaves." I sounded like such a dork. Why hadn't I worn any makeup?

"I can help you with both," said Captain Charming. "And what in the devil's name are you doing on a ship like that, if you don't mind me asking?"

I didn't mind one bit. I was melting, and not from the heat. The office was amply air-conditioned. I explained, flattered by the attention, and startled by how starved I was for civilized male attention.

"Here's the phone," he said, after I'd shared the condensed version of the trip, practicing the script I'd use over and over in the weeks to come once I got home. "Take as long as you need."

I placed a collect call home, fidgeting as it rang and the operator asked if my parents, who I'd clearly woken up, would accept the charges.

"Of course," said my groggy father. No way to perpetrate our Middlebury charade now.

I could just picture them lying in the queen-sized bed that took up most of the floor space in their lake cabin bedroom, covered, even in summer, in a rag-tag assortment of quilts left by my mother's Aunt Miriam and Uncle Eddie. The sun would be refracting through the crystals my mom hung in the large picture window, making rainbows on the brown, braided area rug and maple-syrup-colored wood floors. They would watch the resident swan couple shepherding this season's cygnets around the lake in search of breakfast. All of close to different water.

"Hi, my honey!" said my mother. She must have taken the phone and kicked my father out of bed to grab the olive green extension on the kitchen wall.

"Hi, Klube!" he said, "How are you? We miss you. It's great to hear your voice. When will you be in New York?"

The nickname "Klube" came from a long-forgotten restaurant on 23rd Street that he would pass on his way to work at Metropolitan Life, where he'd worked for twenty-five years. The restaurant had changed hands over the years, but by that time the latest owner had removed layers of paint to reveal the original sign. KLUBE shone through again, dark green lettering outlined in gold painted on glass just over the door. He just liked the sound of the word, and the term of endearment stuck.

My mom seemed to have disappeared momentarily. When he took a breath, she said, "Are you okay, my honey?" So much simpler. One question at a time.

"Yes, I'm good. I'm in Charleston. I didn't know we were stopping here. I'm so glad to be back on the East Coast and to hear your voices. You have no idea." I won't cry, I won't cry, I won't cry. I dug my nails into my left palm to stem the tears.

"Are you going into Charleston?" my dad asked. "It's a great town. Take advantage of being there. See if you can get in. Maybe you can go on a tour, you know, get an overview, a lay of the land." If my mom had been next to him, she would have nudged him or pantomimed for him to zip it, but again, it was hard for her to get a word in.

I wanted to say, *For God's sake, Dad, I've just sailed halfway around the world and back. I think I can skip one city,* but instead

I said, "I'm not sure. We don't have much time here. We will unload very quickly. How is everyone? Is Suzanne there?"

I could hear movement—my mother must have lifted up her elbow. "Suzanne, Suzanne! Your sister is on the phone. She's in South Carolina!" Surely Suzanne was still asleep, and a gentle tap on the shoulder might have been more effective, but it was nice to hear the excitement in my mother's voice.

"How are you, my honey?" she asked again. "You know you can come home earlier if you want, now that you're so close."

"I'm good, Mom. I love you. I'm okay. Thank you, but it will only be a few days more."

"We love you, too. She's okay, Barbara," said my dad. "It will only be a few more days. When will you be in New York?"

"Well, we have Halifax and then Boston after this. It just depends on when we can dock and how long it takes them to unload. If I've learned one thing about ship schedules, it's how unpredictable they are. Maybe Thursday or Friday."

"Thursday would be better. There will be so much traffic on Friday," he said.

As if I could control that. I thought, *Sure, let me just chat with Herr Kapitän Beucking and explain that it'd be more convenient for you if we could arrange to dock on Thursday.* Instead I said, "I have no control over the ship, Dad." He liked to have control over everything. I would not be at all surprised if he called Mr. Williams to ask if he could speed things along. Not so much that he wanted to see me a day earlier. He wanted to manipulate the operations of a gargantuan container ship and four ports to make his ride down Route I78 West smoother.

"I know, I know, honey. I hope you have a chance to get out and see Halifax, too. It's supposed to be beautiful." *I'll add that to my agenda*, I thought.

"Diane!" my sister said. How nice to hear someone use my given name rather than calling me "Fraulein Meyer." I could see her snuggled up under the covers with my mother. With all the windows open, the lake air cooled the cabin and made air conditioning unnecessary. "How is it?" She sounded sleepy, too.

"I'm great. How are you? I'll tell you all about it when I see you in a few days. I can't wait to see you. How were the Navajos?"

"Great. . . ."

"Okay, honey. We can't wait to catch up with you when you get home. Call us as soon as you know when you'll arrive. We love you," said Dad. I knew he didn't want to run up phone charges. Collect calls were expensive.

"I love you guys too. I can't wait to see everyone."

I hung up. I stood for a moment with my hand on the receiver and my back to the guys to make sure I wouldn't cry. There was so much I wanted to tell them. I didn't even know where I'd start.

I could feel the southern beaux staring at my back, curious but not wanting to intrude.

"Thanks so much," I said when I finally turned around.

"Now, do you want to get into town?"

I asked how long it would take and if they knew of any tours. We collectively concluded, and I think that I was a little relieved, that it might be too rushed to try to get in and see much of anything given the unpredictability of our duration in port. I felt a little defiant, too. I was tired, and regardless of what my dad would do, or thought I should do, in my situation, I just wanted to go back and watch the cranes cherry-pick their harvest from the ship. And dare I even think it . . . start to pack? I'd finished the collage and read every book I had. There was not much left for me to do.

Without the pressure of conquering Charleston in a few hours, I lingered and chatted with my crush a bit longer. I had more fun than I'd have had on any tour.

"Well, ma'am, it's been a pleasure speaking with you. We will have the distinct pleasure of escorting you and your ship out of our lovely harbor later this afternoon. We do hope you have the opportunity to return soon, and have a longer visit."

I waved at them from the deck when they, keeping their

word, piloted us out of the port on their tug at around 1600. I felt quite satisfied with what I'd seen of Charleston.

"Come, I want to show you something," said Karl after we'd finished dinner. It was still light as we walked out onto the container deck. We walked aft, to the now mostly empty stern of the ship. When we neared the back, he pointed and laughed a little: "Our 'spare tire'!" I looked down to see what I could not when he'd given me the tour of the belly of the beast that I'd developed such deep respect and affection for. I had no idea, but we had a spare propeller on board. It was, as I wrote in my journal, "humungous." The gleaming, razor-sharp blades could split one of my split ends several times over with ease. I had no idea how they'd switch it out if they had to, but I supposed it would have been easier than awaiting a new one from Germany. Like a guardian angel that I hadn't known had been there, it would have rescued us.

The Atlantic got greener and greener as we made our way north. The Emerald City . . . my ticket home. This water had wrapped me in an ever-constant yet ever-changing womb for this whole trip.

How odd it was to be on this immense ship, and yet to have felt so contained by my cabin, my gender, my language. And to be dwarfed by the unimaginably more immense ocean. It had changed my perspective forever. Life on a floating village, made diminutive by the infinite liquid surrounding it that contains most of the earth's life, at once imprisoned and liberated me. I wondered if I could ever explain this to anyone at home, or if I'd even try.

I thought about the people I'd met in Australia and New Zealand, about all the people I'd left at home. A *landlubber* is someone unfamiliar with the seafaring life, but I understood it as so much more. The globe appeared to me more as an astronaut would see it from space: dots of light strung together showing where much of the population clings to a coastline—to terra firma—to little bits of land and contained lives. With no concept of how minute they are relative to the vastness beyond their shores or their fields of vision. I could understand the once-held belief that there is nothing beyond the edge. Nothing but blue.

"Mrs. Popcorn!" Ingo said, "Halifax today! Maybe you can go find a Canadian boyfriend!" Some of the crew had taken to using his new nickname for me; referring to the substance they claimed filled American boys' muscles. They so amused themselves. I so ignored them.

"But we leave quickly. Maybe two, three hours. You stay close by unless you want to miss your ride home!" He was still very amused with himself.

So much for Halifax.

Boston teased me, too. Rumor held that we'd arrive at 0700, but would reportedly have to anchor and wait for another ship to leave. We were early because things went so quickly at Halifax. I cared for two reasons: I had college friends in Boston, and I longed to sit in Faneuil Hall with them drinking cheap pitchers of bad beer. And our arrival time in Boston, by domino effect, would determine when we'd get to New York. I'd call my parents as soon as we docked. It was almost physically unbearable to be so close to home but so completely unable to get any closer any more quickly.

We anchored outside the harbor at 0830. So close yet so damn far from Boston and from home. I wanted to jump overboard and just swim. It was better on the other side of the world, when home or something close to home wasn't even an option. Then I had to just put it out of my mind. But now it dangled again, like a teasing cat's toy: the Atlantic, the I-95 corridor, American English. I could smell home as surely as the popovers baking for schmoke time.

At 1500 hours we still sat, again. I was tired of playing the Dickens game. The Great Expectations were going to make my head explode. Rumor—again, those tantalizing rumors—had it that the pilot boat would come for us at 1600, but I trusted neither these estimates nor the men who gave them to me. The latter reveled too much in my angst, so I no longer gifted it to them.

We didn't dock until 2000, too late by then to meet up with anyone because we still had to clear customs. When I finally got to the port office I called a few friends. George's line beeped busy, and Randy didn't answer. Then I called home.

"Hi, my honey!" said Mom. "Hold on, I'll get Dad on the line."

"So what's the news, my Klube?" he asked. They sounded very excited.

"We're in Boston. They tell me we will leave here at oh six hundred." God, I was speaking in ship lingo. "At six a.m. So that means around seven Friday morning." I did not apologize for my inability to arrange it so that he'd miss traffic. "But sometimes we have to wait for a spot if another ship is in it, and sometimes the tugboats are busy, and it can take a long time to clear customs and immigration." I was practically panting.

"Listen to you," said Dad. "You old sea dog." I breathed. I laughed.

"I just don't want you to have to wait too long."

"I'll be there at seven, my love." I knew he would. I had inherited my punctuality and impatience genes directly from him. I would hardly be surprised if he drove down at night and slept in the car to avoid traffic and arriving late.

"And, Dad, can I ask you a favor? I'd really like to get something special for my boss, Herr Most, the chief steward. He really likes this German champagne—it's called sekt—the brand is Henkell. If you can't find that, any German sparkling wine will do."

"I'm on it. I'll find it." And I knew he would. I also got my obsessive and tenacious genes from him.

"Do you want to go to the Drop Zone for dinner?" asked my mom. I loved my family. Food first. I couldn't believe I was making dinner plans with them. That I wouldn't be sitting eating brown gravy-covered meatballs with spaetzle and sipping green iced tea. I could hardly contain myself.

"Yes!" The Drop Zone was a very inexpensive, very dark, very mediocre restaurant that served very cliché Italian food—all for a *prix fixe*. We loved it. I almost cried with the wave of craving all things familiar that washed over me.

"Okay, my honey. Can't wait to see you." I felt like Odysseus. But instead of Penelope, my mom awaited my return.

My last night on board was underwhelming. I did treat everyone to a beer, despite Herr Most's protest. "They can buy their own beer," he said. I ignored his grumbling and stuck with the tradition.

I made my rounds to say goodbye to those who I cared about and who had shown me some kindness: Karl, Ingo, Alois, Claudia, Tim . . . and some of the officers. They all wished me well, some with hugs, some with handshakes. We promised to keep in touch, but I knew that to be a polite charade. I got my remaining cash from Herr Stuhlemmer.

They were all busy making their own plans. They stayed a long time in New York relative to other ports. Many would depart the ship with me, to return home to Germany for a while before their next tour of duty. Others had a long shore leave there. My departure was a footnote for them, as had been my tenure on board. Not the monumental monolith that it had been for me.

I finished packing the last few things and waited in my

cabin. The completed, shellacked collage stood propped up on the table: the journey in a nutshell. I doubted that I'd sleep at all, with anticipation caffeinating my system.

Before I locked the door to my cabin for the last time, I knocked tentatively on Herr Kapitän Beucking's door. I was afraid he'd be there; I was afraid he wouldn't.

"*Ya?*"

"It's Fraulein Meyer, Herr Kapitän." He opened the door.

"*Ya?* Is everything okay?"

"*Ya, ya.* I leave tomorrow in New York. I just wanted to say goodbye and thank you in case I don't see you. I really appreciate you letting me spend the time on board, I know it wasn't. . . ."

"*Nein, nein,* please Fraulein. *Nicht.* It was nothing. It was a pleasure to have you travel with us. I hope you have enjoyed our ship." He bowed his head slightly, took my hand in both of his, and shook it firmly.

"Good luck to you."

"*Danke. Gute Nacht.*"

Why was the ocean blue? Was I blue? According to the National Oceanic and Atmospheric Administration, "The ocean is blue because water absorbs colors in the red part of the light spectrum. Like a filter, this leaves behind colors in the blue part of the light spectrum to see." (www.ocean service.noaa.gov)

This ocean had tried to swallow me, too. To drown me in its depth and absorb my colors to prevent my full spectrum of light from shining. At times I felt the tug of the undertow allowing only the blue to rise. But like the ship, her defiant crimson shining above the surface and not succumbing to the pull, I'd managed to stay afloat too.

She wore her red like that eponymous badge of courage, and so, I would too. My colors would radiate like those of the myriad rainbows that rose above the water and graced the horizon during the journey.

Disembarkation

August 17, 1979

40.685649N, 74.07154W

They finally let Dad on board to pack me up and whisk me off after a nearly three-hour customs delay. The gargantuan red *TS Columbus Australia*, on which I'd just spent eleven weeks, dwarfed his British green VW Karman Ghia parked just at the base of the ramp. The green against the red evoked Christmas, even though it was hot, humid mid-August. I hung over the same railing looking down on him that the sailors had hung over watching me the day I arrived.

Herr Most appeared in the hallway just before we disembarked, like Peter Lorrie skulking around Rick's Café American in Casablanca. "Psst," whispered this stoic, stern, buttoned-up boss of mine. I'd been looking for him all morning. "*Auf weidersehen*," he said, as he embraced me. I handed him the Henkell. And wait; was that a tear in the corner of one eye? "Good luck to you, Fraulein Meyer."

And as I prepared to go ashore, onto this dry land I'd longed for since the minute I left it, warm salt water sprung to mine, too. "*Danke, danke*, for everything."

"*Ya, ya*," he said dismissively. "You go now."

"It's so good to see you. I love you, my Klube," Dad said as we sat in the car together, his hand on the clutch-less stick shift. I rested my thumb pad on his flat, concave thumbnail—a reflexive gesture of comfort and affection that we'd shared for as long as he'd called me Klube.

"I love you too, Dad. I'm happy to see you, too"

And I was. Happy to see his wavy dark hair, acne-scarred skin, and tangled teeth again in person. These things that must have filled his adolescence with angst, but they ultimately made him strong. I could relate. My stuff crammed in the compact, trunk-less car, we pulled away from the terminal.

"Let's go out on the sailboat when we get up to the lake!" He said.

"No, Dad, I don't want to go for a sail." I was out of the blue.

Acknowledgments

Above all, to Marcelle Soviero, my muse, midwife, and mama bear—and all my bear cub classmates. Gateless Goddesses Kate Gray and Suzanne Kingsbury. My life support team: Suzanne Fields, Liz Greenberg, Jamie Levine, Randy Kaufman, and Stacey Sobel. Sally Allen for supporting and featuring my writing. And the community at SheWritesPress.

Diane Meyer Lowman's essays have appeared in many publications, including *O, The Oprah Magazine*; *Brain, Child*; *Brevity Blog*; and *When Women Waken*. She also writes a weekly column called *My Life on the Post Road* for Books, Ink (books.hamletlhub.com). She's explored other forms of literary expression in nearly 1,000 haiku poems and many essays about all of Shakespeare's plays. Lowman teaches yoga, provides nutritional counseling, and tutors Spanish. She recently received an MA in Shakespeare Studies at the University of Birmingham's Shakespeare Institute.

Author photo © Devon Lowman

SELECTED TITLES FROM SHE WRITES PRESS

She Writes Press is an independent publishing company founded to serve women writers everywhere.
Visit us at www.shewritespress.com.

Accidental Soldier: A Memoir of Service and Sacrifice in the Israel Defense Forces by Dorit Sasson. $17.95, 978-1-63152-035-8. When nineteen-year-old Dorit Sasson realized she had no choice but to distance herself from her neurotic, worrywart of a mother in order to become her own person, she volunteered for the Israel Defense Forces—and found her path to freedom.

Postcards from the Sky: Adventures of an Aviatrix by Erin Seidemann. $16.95, 978-1-63152-826-2. Erin Seidemann's tales of her her struggles, adventures, and relationships as a woman making her way in a world very much dominated by men: aviation.

Fourteen: A Daughter's Memoir of Adventure, Sailing, and Survival by Leslie Johansen Nack. $16.95, 978-1-63152-941-2. A coming-of-age adventure story about a young girl who comes into her own power, fights back against abuse, becomes an accomplished sailor, and falls in love with the ocean and the natural world.

This is Mexico: Tales of Culture and Other Complications by Carol M. Merchasin. $16.95, 978-1-63152-962-7. Merchasin chronicles her attempts to understand Mexico, her adopted country, through improbable situations and small moments that keep the reader moving between laughter and tears.

Gap Year Girl by Marianne Bohr. $16.95, 978-1-63152-820-0. Thirty-plus years after first backpacking through Europe, Marianne Bohr and her husband leave their lives behind and take off on a yearlong quest for adventure.

Peanut Butter and Naan: Stories of an American Mother in The Far East by Jennifer Magnuson. $16.95, 978-1-63152-911-5. The hilarious tale of what happened when Jennifer Magnuson moved her family of seven from Nashville to India in an effort to shake things up—and got more than she bargained for.